GREEN GUERRILLAS

Environmental Conflicts and Initiatives in Latin America and the Caribbean

A READER

edited by Helen Collinson

Montréal/New York
London

Black Rose Books No. AA246
Hardcover ISBN: 1-55164-067-8 (bound)
Paperback ISBN: 1-55164-066-X (pbk.)
Library of Congress number: 96-79516

Canadian Cataloguing in Publication Data

Green guerrillas: environmental conflicts and initiatives in Latin America and the Caribbean

ISBN 1-55164-067-8 (bound) -
ISBN 1-55164-066-X (pbk.)

1. Environmental protection–Latin America–Citizen participation. 2. Environmental protection–Caribbean Area. Citizenship participation. 3. Environmental degradation–Latin America. 4. Environmental degradation–Caribbean Area. I. Collinson, Helen.

GE199.L29G74 1997 363.7'0098 C96-900773-6

BLACK ROSE BOOKS

C.P. 1258
Succ. Place du Parc
Montréal,Québec
H2W 2R3 Canada

250 Sonwil Drive
Buffalo, New York
14225 USA

99 Wallis Road
London, E9 5LN
England

To order books in North America: (phone) 1-800-565-9523 (fax) 1-800-221-9985
In Europe: (phone) 081-986-4854 (fax) 081-533-5821

Our Web Site address: http://www.web.net/blackrosebooks

A publication of the Institute of Policy Alternatives of Montréal (IPAM)
Printed in Canada

Contents

Acknowledgements

My thanks to my colleagues at LAB: Duncan Green, James Ferguson, Liz Morrell, and Chris Lee, for their guidance; to Emma Pearce for her background research; to Catherine Matheson and Judith Escribano for their translation work; Jutta Blauert for her suggestions on potential contributors; Martin Mowforth for his advice on ecotourism; David Satterthwaite for his help with the Urban Ecowarriors section; Stephanie Williamson for her agricultural expertise; and, finally, my thanks to all the authors who have contributed to this book for their support, patience, labour, and for the inspiration they will surely generate among the readers of *Green Guerrillas*.

Introduction

The title of this book makes a serious point. 'Guerrilla' – derived from the Spanish word, 'guerra' – was originally coined to describe any kind of irregular warfare. Few could deny that Latin America and the Caribbean have witnessed a war on their environments more dramatic, more widespread, and certainly as traumatic as any military conflict. In the 1980s it was a war that pricked the conscience of the world, with graphic TV coverage of Amazonia on fire, of pristine rainforest gouged out by mega mining complexes, dam-building projects and chainsaws. But media interest in any war is usually shortlived. What has happened to Latin America's environment since then is yesterday's story, unless you are on the frontline. Visit any city, town or rural area of Latin America in the 1990s and you will find conflicts over the control and use of natural resources, over the health risks of pesticides, industrial pollutants and poor sanitation; disputes triggered by the dumping of toxic wastes, by the granting of mining rights or by major industrial development plans. In Amazonia, too, the war goes on. In January 1996 the Brazilian government announced Decree 1775, opening up 57 per cent of all indigenous land in Amazonia to claims from outside interests, and unleashing a renewed invasion by loggers, ranchers and miners of this fragile ecosystem.[1]

Aside from the conflicts themselves, the title, *Green Guerrillas,* also refers to Latin American environmental campaigners – the continent's 'irregular warriors'. Paramilitary guerrillas they are not, though some have died for their cause. In December 1988 the assassination of prominent Brazilian forest campaigner Chico Mendes sent shock waves through the environmental movement worldwide, as Anthony Hall describes in this book. In fact, the only unique feature of Chico's case was his international profile. The recent murders of green activists in Honduras, Colombia, and even Costa Rica (a country which prides itself on its environmental awareness)[2] are reminders that environmental campaigns can strike at the heart of political and corporate power. The hostility some campaigners have provoked often mirrors attitudes towards the continent's traditional left. 'They call me a green communist' claims Salvadorean environmentalist, Ricardo Navarro.[3] One reason why Latin American environmental campaigners are more vulnerable than their fellow activists in Europe or North America is that Latin America's democratic and judicial institutions are still weak and protesters often have limited recourse to the law.

But *Green Guerrillas* is not about these relatively isolated acts of violence or their perpetrators. It is more concerned with how Latin Americans are combatting environmental degradation in their daily lives and the extent to which they are managing to influence state and corporate policies. Aimed at a broad, non-specialist audience, the book will be of particular interest to environmental campaigners, undergraduate students, and visitors to Latin America.

Contrary to the doom and gloom scenario we have come to expect, there are many success stories, some of which are related in this book. Marianne Meyn reports on the successful grassroots campaigns to change Puerto Rico's energy policy; Peter Rosset welcomes the nationwide adoption of organic farming in Cuba; Stephen Nugent and Catherine Matheson highlight the progress made by sustainable agro-forestry projects in Brazilian Amazonia. Even where specific campaigns are not won, Jorge Cela from the Alternative City project in Santo Domingo points out that the environmental awareness generated in the process of taking action can be just as important as the original goal. It is encouraging that local ecological committees, green graffitti, and environmental articles in the press have now become habitual features of Latin American cultural life.

In the opening 'Continental Perspectives' section, David Kaimowitz shows that while some of the pressure for environmental reform in Latin America has come from the environmental movement internationally, rarely have Latin American governments taken action in defence of the environment without pressure from below, from their own citizens. If we accept his assertion, then it is essential to identify *who* from below is exerting this pressure and *why*. For that reason, *Green Guerrillas* focuses far more on Latin American people than it does on the environment itself.

Many of the book's contributions emphasize the extent to which Latin Americans' concern for the environment is motivated by basic and immediate material interests. In some cases, environmental destruction can wipe out livelihoods and whole communities. Denise Stanley explains how Honduran fishermen are opposing the clearing of wetlands for shrimp farming because it is perceived to endanger local fish stocks and restricts their access to coastal areas; Charles Arthur informs us that Haitian peasants are taking action on soil erosion because their meagre plots yield less and less crops; Aidan Rankin's profile of the Wichí people in Argentina describes how an invasion by cattle ranchers (which has turned the Chaco region into a dustbowl) threatens the Wichí's very existence: 'our land is dead and we are dying of hunger.' On a more positive note, Margarita Pacheco reminds us that for the poor of Bogotá, recycling was a livelihood and a source of income long before the rise of environmentalism.

Not that the poor and the marginalized are the only green guerrillas in Latin America. Compared to its counterpart in the North, Latin America's environmental movement is extremely heterogeneous. Contributions by Hilary Beckles and Nick Caistor demonstrate that the recipe for a successful environmental campaign is often collaboration between marginalized sectors on the one hand and on the other, middle-class, (usually urban-based) environmental campaigners with the time and money to devote to campaigning.

The predominance of material motivations may mean that Latin American communities' relationships with their environments are not always as ecologically pure as environmentalists in the North might wish. Anthony Bebbington's study of the Ecuadorean Andes, for example, illustrates the extent to which Indian farmers' methods are determined by economic and social pragmatism. Despite the perception that indigenous peoples are the guardians of a purer, more sustainable, traditional agriculture, Bebbington claims that the farmers in his study willingly discarded traditional practices and embraced modern farming technologies (including the use of chemical inputs) where the latter were seen to be more beneficial. Similarly, Lucy Alexander stresses that the neglected inhabitants of Colombia's Pacific coast are not averse to the development of the region *per se* if it will improve their quality of life, while Janet Townsend dismisses ecofeminist claims that women have an inherently ecological attitude towards the environment. Her research on women settlers in Mexico's Lacandon forest suggests that women may be just as keen to cut down trees as men, where it represents the only solution to the family's economic difficulties.

The common denominator behind many of the conflicts cited in this book is ordinary people's desire to have some control over the development of their environments. The struggle for democracy is therefore linked to the environmental cause. Another common concern is that development should be sustainable. There is a growing recognition that if development is not *environmentally* sustainable, then in the long-term local communities will not benefit from it.

The distinction between goodies and baddies in Latin American environmental struggles can be very blurred. Marginalized peasant communities have been held responsible for much of the continent's deforestation in recent decades (because they have had nowhere else to go *but* the forest). On the other hand, case studies in this book from Haiti, Mexico, Brazil, and Ecuador suggest that yesterday's plunderer can be tomorrow's saviour. Stephen Nugent points out that because forest colonizers have to deal with the negative effects of deforestation on a daily basis, they may be the ones most eager to rectify the problem. At a macro level, it is ironic that in 1996 environmental campaigners are appealing to the World Bank to intervene on their behalf against the Brazilian Decree mentioned above.[4] Ten years ago, the World Bank – with its awesome development plans for Amazonia – was environmentalists' enemy number one. Have the tables turned? In the book's opening piece, Elizabeth Dore describes the 'greening of the discourse', prompted by the international environmental movements of the 1980s. By 1992 the World Bank had decided that 'working for a better environment' was its major policy goal.[5]

It remains to be seen whether the World Bank's green rhetoric will be matched with real action on Brazilian Amazonia. Elizabeth Dore is sceptical, pointing to the World Bank's energy policy, amongst others, which still gives priority to large dams and conventional power plants over alternative energy policies. She asserts that the fashionable rhetoric about environmental awareness and sustainable development is meaningless and contradictory, unless macro-economic policies are altered.

If there *is* a single enemy confronting the Latin American environment, the free market model of development is a leading candidate. On the one hand, there has been a greening of the discourse; on the other hand, there has been a voracious application of market economics across Latin America in recent years, with serious consequences for the environment, as several contributors point out. While Latin American governments must take part of the blame for the war on Latin America's environment, few contributors to this book ignore the pivotal role played by international financial institutions, multinational companies, and other powerful forces outside Latin America, in promoting the free market model.

Since few Latin American countries have a developed manufacturing base, they have traditionally relied on the export of agricultural commodities and natural resources. With the rise of market economics in the 1980s, this export drive has been intensified. The abuse of pesticides highlighted by Sarah Stewart and Lori Ann Thrupp, for example, is seen as a consequence of the pressure to expand agribusiness into non-traditional export products. The extraction of oil, minerals, timber or forest plants from the Ecuadorean Amazon or the Colombian Pacific region (see Judith Kimerling and Lucy Alexander in section 2) is part of the free market rationale that undeveloped regions should be opened up and their riches plied on the international market. In addition to the physical transfer of natural products, the countries of Latin America and the Caribbean are being marketed as an increasingly popular tourist destination. In the sixties and seventies, only the rich and famous took holidays in Mexico, Rio, or the Caribbean. The advent of mass tourism has reversed this trend, but with a heavy environmental toll, as Polly Pattullo and Hilary Beckles illustrate. In those regions where industrialization has occurred, effective environmental controls can significantly reduce pollution (as demonstrated by Julio Dávila and Jonas Rabinovitch) but deregulation and privatization (hallmarks of market economics) are undermining environmental controls. Marianne Meyn asserts that pharmaceutical companies were lured to Puerto Rico, for example, in part by less rigorous environmental regulations than on the US mainland.

Several contributors question whether such a radical and rapid integration of the continent into the world market is really necessary or whether it can be achieved without the wholesale destruction of its environment. While opinion is divided, all contributors do believe there are viable and realistic alternatives to the dominant model of development and mode of practice. In fact, out of every conflict cited in this book, concrete proposals have arisen for doing things differently, many of which are already being implemented.

Some of the most exciting initiatives seem to be taking place in cities. This is perhaps not surprising, as Julio Dávila reminds us that the majority of Latin Americans are now urban. A massive influx of rural migrants over recent decades has put extreme pressure on the environmental health of most cities. Invariably, it is the newcomers themselves who have confronted the environmental problems, as they are the ones most affected by them. With few resources of their own, a key aim of many urban campaigns is to extract concrete commitments from the state – a hard task made harder by shrinking state revenues (see Jorge Cela's piece on Santo Domingo). On the other hand, the pressure is mounting for municipal authorities to take more radical action. Julio Dávila believes that city governments stand at the core of a democratization process which has swept through Latin America after predominantly military rule in the 1970s. Effective decentralization programmes and local elections are poised to increase public participation in the running of cities. The case of Curitiba in Brazil (profiled in the final essay) offers a rare glimpse of the environmental revolutions municipal authorities could bring about if they were to take a more proactive approach. 'What makes Curitiba unusual,' explains former city planner Jonas Rabinovitch, 'is not so much that it had a coherent plan, but that it was implemented.'

Returning to the original use of guerrilla as a word for irregular warfare, 'irregular' seems a highly appropriate description of the varied and innovative tactics adopted by Latin Americans to bring the war on their environments to an end. True, the general prognosis still looks poor. But the case studies in this book show that for many of those Latin American and Caribbean communities who are confronting environmental degradation face-to-face in their respective localities, all is not lost.

Helen Collinson
March 1996

[1] Jan Rocha, 'Law unto himself', *Guardian*, 27 March 1996.
[2] Luis Erasmo, leader of environmental community action in Milan, Caquetá, Colombia, murdered in October 1995; Jeannette Kawas, President of the Honduran environmental organisation, PROLENSATE, murdered in February 1995; David Maradiaga, leader of Friends of the Earth Costa Rica, died in suspicious circumstances in August 1995, while coordinating a campaign against a gold mining company.
[3] Ricardo Navarro, quoted from a lecture he gave at the Institute of Latin American Studies, London, May 1992.
[4] Jan Rocha, op cit.
[5] World Bank, *World Development Report 1992*.

1
THE CRISIS AND THE MOVEMENT: CONTINENTAL PERSPECTIVES

Capitalism and ecological crisis: legacy of the 1980s

Elizabeth Dore

Since the 1980s the ecology question has occupied a prominent position in Latin American political debates. This reflects the importance of the global environmental movement as much as regional preoccupation with environmental conditions. Discussion of ecological issues centres on two questions: how to achieve sustainable development, and how imperialism conditions environmental change. In practice, however, the two approaches overlap to form a rich, if sometimes eclectic, debate.

Radical environmentalism in Latin America

In the 1960s and 1970s progressive Latin Americans embraced dependency theory to explain underdevelopment. They held that Latin American countries were poor because advanced industrial countries appropriated their economic surplus. Leading Latin American environmentalists have elaborated a theory of eco-dependency in which surplus extraction takes the form of ecological pillage. They argue that companies earn a higher rate of profit in Latin America than in the US or Western Europe because there they can despoil the environment far more than would be tolerated 'at home'. This devastates the region's resource base and reduces long-term potential for capital accumulation.[1] A variant of eco-dependency is that underdevelopment is caused by the use of 'foreign' technologies. In this view, European methods of production are incompatible with the tropical climate and soils found throughout Latin America. Over time inappropriate technologies degraded the fragile ecosystems of the region. For example, cattle-grazing and mono-crop agriculture are sustainable in Europe and the United States but not in Latin America, where they were introduced after the Conquest causing long term, irreversible erosion.[2] An explicitly anti-modern strand of Latin American radical environmentalism stresses the sustainability of indigenous production systems, and the unsustainability of European ones. Writers of this persuasion hold that small scale production is inherently good, and large scale bad; that traditional technologies are always green, but never modern industry.[3]

There are compelling aspects of this anti-modern critique. Yet, judgements about the environmental impact of production techniques when they are divorced from analysis of the social system are problematic. Small-scale production may or may not be ecologically preferable to large-scale. The central issue is not the size of the production unit, but who controls it and for what purposes. In so far as Indian and peasant communities practice more sustainable forms of natural resource use it is because they are not driven by capitalist competition. The challenge is not to do away with industrialization, but to de-link industrialization from profitmaking.

Sustainable development

In the 1980s sustainable development became a catch-all term widely used to frame environmental issues. It has come to mean 'development that meets the needs of the present without compromising the ability of future generations to meet their own needs'.[4] As this definition is both malleable and moral, sustainable development has become the broad church of environmentalism. Advocates of sustainable development offer an array of proposals, from liberalizing markets to expanding the power of the state, concentrating on how to manage the economic system to make it less polluting and less wasteful of resources. The concept is particularly favoured by policymakers because it obscures ideological differences about the causes of ecological destruction and what should be done about them. A hallmark of this eclectic approach is the assumption that countries face a trade-off between growth and environmental degradation. Sustainable development is not about transforming capitalism; it is about reforming the status quo to make capitalism more eco-friendly.

Despite its reformist character, sustainable development marks a significant advance in environmental politics in Latin America. Historically ecology movements in the region campaigned to protect any endangered species except endangered homo-sapiens. Environmentalists lobbied to preserve rainforests, watersheds, birds and animals, but not people. Instead of targeting the system that drives the poor to overexploit their environment, often ecologists saw poor people as the enemy. Sustainable development made the environment into a human issue by linking the protection of people and ecosystems.

Two problems with the sustainable development approach stand out. First, while tropical rainforests, especially Amazonia, monopolise the attention and funds of advocates of sustainable development, the vast majority of Latin Americans live in areas which attract minimal international interest. Second, the notion that economic growth inevitably causes ecological destruction prevents ecologists and policymakers from exploring the connections between property relations, political power and environmental destruction.

In the 1980s the multilateral lending agencies adopted sustainable development. Institutionalization turned it into the latest in a long parade of development fads: growth with equity, integrated rural development, basic needs, appropriate technology, and women in development. Each began as a critique of the orthodoxy of the development establishment. Then each was appropriated by those agencies as a way of defusing opposition, leaving the traditional agenda of the development community virtually unchanged. Sustainable development suffered the same fate.[5]

An environmental critique of development practice briefly altered the rhetoric and bureaucracies of the major lending agencies. For example, in its 1992 World Development Report the World Bank announced that 'working for a better environment' was its major policy goal. Quickly sustainability rose to a prominent place within the development establishment.[6] Speeches and working papers heralded the new policy. New programme departments and professional specializations created vested interests in sustaining the fashion. The World Bank appeared to have assumed the mantle of environmentalism, but it was form over essence. Environmentalism made little impact on the Bank's core activities: adjustment lending and project aid. Despite high profile pronouncements about the importance of sustainability, concrete measures to improve the environmental impact of the Bank's activities were negligible. In the energy sector, where specialists agree that huge power schemes are neither ecologically nor economically sustainable, the World Bank continues to fund large dams and conventional power plants instead of alternative energy policies.[7]

In the end, environmental conditionality in project aid proved less significant than the ecological impact of the neo-liberal economic policies that lending agencies fostered.[8] The International Monetary Fund and the World Bank imposed adjustment programmes designed to ensure that countries increased their foreign exchange earnings to service the debts they accumulated in the previous decade.[9] To this end, the agencies required that countries increase their trade balances, which meant promoting exports. This was disastrous for poor people and for the environment.

To comply with the agencies' ground rules, governments sought private investment to expand exports. As Latin American countries simultaneously were trying to attract capital, they competed over which could provide the most favourable conditions which included, among other things, low wages, a submissive labour force and few environmental restrictions. The policy was successful on its own terms. Exports of natural resources rose spectacularly. In the 1980s the value of Latin American fishery exports was almost four times greater in the previous decade. Forestry exports rose as well, and the production of iron and copper was higher than in the 1970s.[10] Exports of selected agricultural products also grew significantly over the same period.[11] A few countries in the region succeeded in attracting low wage, high pollution industries, most notably Mexico. With pressure to export and generally to comply with IMF and World Bank economic conditionality, Latin American governments implemented development strategies which were antithetical to environmental sustainability.

Natural resource destruction accompanied the drive to export; damage was compounded by the economic orthodoxy of the development establishment. Free market economics became an article of faith in the multilateral lending agencies in the 1980s, as in the economics profession at large. Free-marketeers argue that private sector cost-benefit calculations will result in environmental improvements, that the costs to capital of ecological deterioration and natural resource depletion will compel firms to reduce pollution and take measures to conserve ecosystems. In this world view, market liberalization, free enterprise and reduction in the role of the state will improve ecological conditions.[12]

But contrary to World Bank statements, the environmental impact of liberalization was not positive in Latin America. Weak environmental regulations encouraged entrepreneurs to adopt production methods which ensured large profits in the short term with little regard for natural resource preservation. Industries were attracted to Latin America because they could pollute, dispose of toxic wastes, and extract resources with minimal state interference. Where governments introduced environmental regulations, often responding to pressure from the lending agencies themselves, their monitoring capacity was negligible. With pared budgets and reduced staffs, state agencies charged with implementing environmental regulations were unable to enforce their mandate. Environmentalism became yet another avenue for market liberalization and reduction in the ambit of the state.

Few professionals challenged the way the development establishment framed environmental debates. To retain credibility and access to the corridors of power, environmentalists raised issues of sustainability in terms acceptable to the agencies. Ironically, instead of the environment movement transforming the priorities of the multilateral agencies, the reverse occurred. The agenda of the World Bank and the IMF – export promotion, free markets, a small state – became central tenets in the ecology debate.[13]

By the 1990s it was clear to progressives in the environmental movement that the international development establishment had wrapped itself in the rhetoric of environmentalism to legitimate a structural adjustment agenda that was a barrier to sustainability and human development. As well as directly fomenting irrational resource use, structural adjustment intensified poverty throughout Latin America. In

1980, 35 per cent of the population of the region lived below the poverty line. Ten years later the figure was 50 per cent.[14] Poverty is intensifying pressure on the environment in numerous ways. In the countryside people are cultivating land that had been considered too eroded for agriculture, accelerating soil depletion. In urban areas reductions in subsidies and incomes have put the price of fuel beyond many people's means. As a result they cut firewood in zones surrounding cities, denuding those regions of trees.

Socialist environmentalism

The motor force of capitalism is the production of profits. What to produce, how to organize production, and how to distribute commodities are determined neither by social needs nor ecological sustainability, but by profitability. As growth is key to raising profits, the system continually engenders the expansion of production, consumption – and of waste.[15]

While ecological destruction is inherent under capitalism, there is scope for environmental reform. In the past decade environmental movements have forced politicians to address issues of ecological change. Local resistance to extreme contamination and natural resource destruction is common in Latin America. wary of denunciations by national and global groups, governments of the region are beginning to pay attention to the environmental impact in their policies. Sometimes this has resulted in curbing ecological damage and its social effects. Frequently, however, official attention takes the form of paying lip service to natural resource protection. The track record of Latin American governments on environment issues is not good. Bowing to pressures from national and international capital, official policies that protect profits usually take precedence over those that protect ecological sustainability, as demonstrated by the following case studies of Amazonia and Mexico City.

Amazonia

The bread and butter of popular environmentalism in the United States and Europe is the myth of the destruction of the Amazon. It is a tale of rapid, irreversible deforestation and species extinction, and of global warming. While not without truth, this emphasis hides the social and political struggles that will determine the fate of the forest and its fifteen million people.[16]

For more than a century competing groups have fought over the Brazilian Amazon. Each had a different vision of how the natural resources of the region should be used.[17] A volatile mix of farmers, ranchers, miners, land speculators, indigenous tribes, rubber tappers, politicians, multinational companies and international banks all have laid claim to the resources of the region. In the last ten years the terms of those struggles have changed dramatically. In 1970 the debate was about how to modernize the region: how to turn the 'empty' forest into a productive resource. By 1990 the discourse was green; all claimants cast their demands and interests in an environmental framework. Nevertheless, optimism may be premature. Environmental sustainability and social justice are key words in the debates on the future of Amazonia. But deforestation and social violence continue. Subordinate groups who loudly press their claims are intimidated, repressed and silenced much as before.

In the early 1970s the military regime in Brazil embarked on an ambitious project to open up Amazonia. The rainforest, like tropical forests elsewhere, was seen as a vast empty space which called out for development. Development meant building roads, bringing in settlers, promoting economic activities, and taking out products. At that time all sides in the public debate over the future of the region favoured clearing the forest. One goal of the first scheme to develop the Amazon was to resettle landless families, particularly from the impoverished Northeast, on plots of 100 hectares. Official propaganda declared that the advancing Transamazonica highway would relieve problems of poverty and landlessness; it would connect 'people with no land to a land with no people'.[18]

As the road advanced into the jungle, people settled along the highway and cut down trees to stake their claim. Land titles and loans were available to settlers who could show that they were using land productively. The easiest way to do this was to graze cattle. Then the early populist development model fell from favour in Brasilia and government policy changed, clearly favouring large private enterprise in the modernization of Amazonia. Within a few years large investors and land speculators began crowding out the settlers. Lured by the tax rebates, subsidies and soft loans originally meant to encourage small farmers to colonize the region, finance capital from southern Brazil invested in landed property. When inflation soared in the 1970s and 1980s, and investors clamoured to put their money into safe assets such as land and gold, the property market in Amazonia thrived. As large capital entered the region, first wave settlers either moved to the edge of the frontier, clearing new lands for themselves, sought work as wage labourers on larger ranches, or returned from whence they came, often poorer than before.

Cattle-grazing created an environmental and a social disaster in Amazonia, as it did throughout Latin America. The infamous burning of the forest, the first step in clearing land, had global and local implications.

Although there is no scientific consensus that reduction of the tropical rainforest produces global warming, forest burning produces considerable quantities of carbon. This heats the earth's atmosphere and contributes to the 'greenhouse effect'. Locally, clearing the land of trees and plants for grazing initiated rapid decline in soil fertility. Even after it became known that clearing the forest would turn the area into a waste land within five to ten years, government policy and financial incentives continued to promote the expansion of pasture land.

Profits in ranching did not derive primarily from beef production. The cattle industry became a vehicle for entrepreneurs to acquire land, subsidized credit and tax reductions. Therefore, destruction of the natural resource base of the industry did not constrain profitability. Rather, it accelerated speculation in land and capital accumulation. In short, there were few economic incentives to exploit the ecosystem in a sustainable way. Quite the contrary, profitmaking promoted the wanton and unsustainable use of land. The abuse of natural resources in the region was so flagrant that it gave cattle raising a bad name, but not because banks and ranchers made a rational appraisal of the economic, social and ecological costs and benefits of the industry. It was communities in Amazonia opposed to ranching and international environmental organizations who exposed the consequences of cattle grazing.

Ranchers and small farmers were joined by others who competed to claim the valuable resources of the forest. Mining was second to cattle in transforming the region. Brazil's Grande Carajás Project, begun in the 1980s, is the largest mining complex in the world. Everything about Carajás was mammoth, including its potential to alter the region's ecosystem. The project converted one-quarter of Amazonia into the world's largest industrial and agro-livestock centre.[19] At the hub of the enterprise are massive deposits of iron ore and manganese which reputedly enjoy the lowest production costs in the world. It was estimated that Carajás alone could produce ten per cent of the world's supply of iron ore.

Grande Carajás became the heart of a vast integrated development project that includes a string of open-cast mines which produce bauxite, copper, chrome, nickel, tungsten, cassiterite and gold, in addition to iron ore and manganese. Radiating from the mines are processing plants, steel and aluminium mills, agro-livestock enterprises, hydro-electric dams, railroads and deep-water river ports. Together these cover an area of 900,000 square kilometres, the size of France and Britain combined. The complex served as a giant magnet, drawing farmers, ranchers, gold prospectors, and enterprises of all kinds into Amazonia.

The project has serious ecological implications, which, because of its scale, were difficult to predict. Soon after building began at Carajás, significant ecological changes became evident. The project involves massive deforestation. 1.6 million acres of timber was cut annually to stoke the pig-iron smelters and provide lumber for construction. The state-owned company which runs the project pays lip-service to conservation and implemented some reforestation. Nevertheless, large areas of the tropical forest were reduced to scrub land. By the time the first iron ore rolled out of Carajás in 1985, rapid deforestation had altered the climate. Less rainfall, combined with soil erosion, siltation and flooding of the region's rivers caused widespread desertification. Extinction of plant and animal species signalled great changes in the ecosystem.

Large as they were, Carajás and other corporate ventures did not monopolize mining in Amazonia. After a steep rise in the price of gold in 1979, *garimpeiros*, or small-scale miners, flocked to Amazonia. The roads, railroads, and services installed for Carajás and for the cattle industry facilitated migration. It is estimated that almost one million garimpeiros panning in and around the rivers of Amazonia produced more than 90 per cent of Brazil's annual output of gold.

From the beginning grazing and mining encroached on lands used by indigenous tribes and rubber tappers. At first their rights to the resources of the forest were ignored because politicians dismissed them as primitives, soon to be swept away by progress. By the mid-1980s the struggle over the fate of the forest appeared, at least on the surface, considerably different. Indigenous groups and tappers were considered legitimate participants in the debate.[20] Their persistent resistance to expropriation and to felling the forest, combined with their links to national and international organizations, converted them from pariahs to legitimate actors in the unfolding drama.[21]

In 1970 there was no legitimate voice calling for preservation of the forest. In 1990 no group that sought legitimacy could oppose it in words. Even ranchers and loggers had to recast their claims in environmental terms. This has been called the greening of the discourse.[22] In 1970 cattle ranching was almost universally promoted as the best way to develop the Amazon. In 1990 it was a symbol of destruction. The struggles of a constellation of forces, led by rubber tappers and indigenous tribes, to claim their rights to the resources of the forest shifted consciousness and politics. That they were moderately successful reflected the influence of environmental movements world-wide over the development debate. Nevertheless, despite the greening of the discourse, and the creation of extractive and indigenous reserves (see Anthony Hall), the main thrust of Brazilian policy in the Amazon continues to promote large capitalist enterprises. Sustainable resource use remains a low political priority.

Urban contamination and corporativist mobilization: the case of Mexico City

By the end of the twentieth century more than seventy per cent of Latin Americans lived in cities. Nevertheless, problems of urban environmental degradation tend to be ignored by ecologists in the United States and Europe. They do not lend themselves to the simple morality of saving the rainforest.

Mexico City is infamous for air pollution. At an altitude of 2,300 metres above sea level, sprawling across a valley of 9,600 square metres, surrounded by mountains on two sides, and occupying the salty, dry beds of what once were a series of lakes, Mexico City is poorly placed for an industrial metropolis. The zone is prone to thermal inversions where lack of wind allows contaminants to accumulate in the atmosphere. Despite these geographic peculiarities, Mexico City's severe environmental problems are little different from those of other Latin American cities.

The capitalist transformation of agriculture, which accelerated in Latin America in the 1950s, set in motion a process of rural-urban migration. Facing a dual process of concentration of land in large capitalist farms and decline in the viability of small-scale production, many peasants abandoned their land; others were evicted. With the Green Revolution machinery, pesticides and fertilizers transformed agricultural production and processing, so only a small proportion of the rural poor found permanent work in the countryside. To survive people migrated to urban centres in search of work. The capital cities of Latin America, with their concentration of industry, commerce and government, became magnets for the dispossessed rural poor. Mexico City exemplified that process. Its population doubled in each decade from 1940 to 1970, a trend that was repeated in many metropolitan cities in the region.

The 1960s and 1970s, the decades that witnessed the urban demographic boom in Latin America, were also decades of industrial expansion. Subsidies and tax incentives promoted the spatial concentration of industry in the region's capitals. Whereas the pattern of urban environmental decay in Mexico City was similar to, though greater than, that found in many Latin American cities, the response of the Mexican government was unique. In keeping with its corporativist character, the Partido Revolucionario Institucional (PRI) created an ecology movement organically linked to the party and the state.[23] In the 1980s growth of the political opposition made it evident that the PRI was losing its legitimacy. One issue that focused discontent, especially among the urban middle classes, was the deplorable quality of life in the capital. The government perceived that opponents might channel this disaffection into a broad movement that could threaten the party's control. To prevent this the PRI launched an environmental campaign to reconquer the political initiative. Party leaders perceived that ecology could be a safe and at the same time popular issue which would enhance the PRI's national and international image.

By commiting the government to improving the environment, the PRI established ecology as a legitimate area of debate. It created a network of supposedly grassroots groups and encouraged their activities. As usual, the PRI used this network to channel rewards, financial and otherwise, to people who emerged as leaders, and to inform on opposition activities. In short, the PRI sought to turn critics into clients of the state. The PRI had some momentary successes nonetheless. Its politics of mobilizing ecology groups undermined the influence of independent organizations that were calling attention to the PRI's obeisance to private enterprise at the cost of environmental degradation.

The government found it could not always control the movement it created. The PRI's pledge to improve the environment, especially in the capital, clashed with the government's neo-liberal economic strategy. In practice, market liberalization and fiscal austerity took precedence over environmental reform. The government encouraged investment by removing barriers to profitability and was loath to impose environmental regulations which might be viewed negatively by the private sector. In addition, by reducing the government budget it curtailed the state's regulatory capacity. As a result, the government was less able to enforce those environmental standards it legislated.

When the government reneged on its public commitment to environmental reform, tensions mounted within the environmental movement. It became apparent that the Mexican government was prepared to undertake only token reforms to alter patterns of natural resource use and abuse. When reform implied a cost to capital or to the state, as they inevitably did, the PRI shed its green veneer.

What might be

Environmentalism has emerged as one of the main forms of opposition politics in Latin America at the end of the millenium. As in other parts of the world, ecology struggles in the region have tended to concentrate on ameliorating particular examples of environmental degradation, rather than confronting its social causes. Ecological disasters in the former socialist countries of Eastern Europe demonstrated that socialism can be as environmentally devastating as capitalism. Radical transformation in the relationship between society and the environment requires conscious critiques of current norms in natural resource use, as well as of the dominant relations of property. The ecology movement has facilitated the former by creating mass consciousness of the consequences of environmental destruction. If that were joined to an ecological critique of capitalism it could provide the framework for a powerful movement for social and environmental change.

[1] Enrique Leff, 'Estudios Sobre Ecología y Capital,' *Estudios Sociales Centroamericanos*, 49 (enero-abril 1989) pp.49-78; *Ecología y Capital: Hacia una Perspectiva Ambiental del Desarrollo*, Universidad Nacional Autónoma de México, 1986. For a Marxist critique of dependency theory see Elizabeth Dore, *The Peruvian Mining Industry: Growth, Stagnation and Crisis*, Westview Press, Boulder, 1988, Cap 1. For a more sympathetic account of dependency theory see Cristóbal Kay, *Latin American theories of Development and Underdevelopment*, Routledge, New York, 1989.

[2] Stéfano Varese develops a variation of this theme. He argues that technologies developed for and by foreign cultures were implanted in Latin America with the Conquest. Therefore, he calls the use of these inappropriate technologies 'internal colonialism' as they are no longer foreign to the region. Cited in Fernando Mires, *El discurso de la naturaleza: ecología política en América Latina*, DEI, San José, Costa Rica, pp.65.

[3] Víctor Toledo, 'Utopía y Naturaleza: El Nuevo Movimiento Ecológico de los Campesinos e Indígenas de América Latina' *Nueva Sociedad*, 122, Nov-Dec 1992, pp.72-85; Stéfano Varese cited in Fernando Mires 1990 , op cit. For a critique of this position see Elizabeth Dore, 'Una interpretación socio-ecológica de la historia minera latinoamericana' *Ecología Política*, 6 April 1994.

[4] World Commission on Environment and Development (WCED), *Our Common Future*, Oxford University Press, New York, 1987.

[5] Ann P. Hawkins and Frederick H. Buttel, *The Political Economy of 'Sustainable Development'* n.d., ms.

[6] The World Bank, the International Monetary Fund, the Inter-American Development Bank and the United States Agency for International Development form the heart of the development establishment.

[7] Bruce Rich, *Mortgaging the Earth: The World Bank, Environmental Impoverishment and the Crisis of Development*, Beacon Press, Boston, 1994.

[8] John Weeks, 'Contemporary Latin American Economies: Neo-Liberal Reconstruction' in Sandor Halbsky and Richard L. Harris (eds) *Capital, Power and Inequality in Latin America* Westview Press, Boulder, 1995.

[9] Dore 1992, op cit, pp.79-85.

[10] Food and Agriculture Organization (FAO) *Country Tables 1989: Basic Data on the Agricultural Sector*, p.327, Rome 1989; Ministerio de Obras Públicas y Urbanismo (MOPU), *Desarrollo y Medio Ambiente en América Latina y el Caribe*, Mexico, 1990.

[11] FAO 1989, op cit.

[12] The World Bank, *World Development Report 1992*, Oxford University Press, New York, 1992, p.10.

[13] Hawkins and Buttel, op cit, pp.4-5.

[14] MOPU, 1990, op cit, p.157.

[15] This gives rise to a tension in the capitalist system: the production of profits undermines the sustainable reproduction of the physical world, which itself is necessary for the continued expansion of capital. See John Weeks, *Capital and Exploitation*, Princeton University Press, Princeton, N.J. 1981.
[16] D. Faber, *Environment Under Fire: Imperialism and the Ecological Crisis in Central America* Monthly Review Press, New York 1993; Peter Utting, *Trees, People and Power: Social Dimension and Forest Protection in Central America*, Earthscan, London 1993.
[17] Much of the argument that follows is drawn from Marianne Schmink and Charles H. Wood, *Contested Frontiers in Amazonia*, Columbia University Press, New York 1992.
[18].Schmink and Wood, 1992, op cit, p.2.
[19] Anthony Hall, *Developing Amazonia: Deforestation and Social Conflict in Brazil's Carajás*, p.20, Manchester University Press, 1989.
[20] The plight of the Yanomami Indians of Brazil received particular attention from survival International. For the impact of mining on the Indians of Brazil see D. Treece, *Bound in Misery and Iron: The Impact of the Grande Carajás Programme on the Indians of Brazil*, Survival International, London, 1987.
[21] Hall, 1989, op cit.
[22] Schmink and Wood, 1992, op cit, p.16.
[23] This argument is inconsistent with that presented in Stephen P. Mumme, 'System Maintenance and Environmental Reform in Mexico' *Latin American Perspectives* 19:1, Winter 1992, pp.123-124; and in David Barkin, 'State Control of the Environment: Politics and Degradation in Mexico' *Capitalism, Nature, Socialism* February 1991, 2, pp.86-108. It also reflects prevailing wisdom within that part of the ecology movement in Mexico which considers itself in the left opposition.

Social pressure for environmental reform in Latin America

David Kaimowitz[1]

The current pattern of development in Latin America is unsustainable. It involves the use and degradation of natural resources at a rate faster than they can be produced and/or replaced. Each year there are fewer forests, minerals, petroleum deposits, genetic resources and fertile soils. Access to water is now a serious problem; cash-crop farming based on the cultivation of a single crop is increasing the continent's vulnerability to pests. The nature of industry, agriculture, transport, and human settlement is causing air and water pollution, which in turn damages human health and the environment.

Solving the problems associated with excessive exploitation of natural resources and contamination of the environment requires state intervention: the creation of a legal and macro-economic framework, a system of taxes, fines and subsidies that reduce the private incentives to contaminate and over-exploit natural resources, investments in technology, education and infrastructure, and the creation of mechanisms for resolving conflicts and disagreements between diverse interest groups. A great deal can be accomplished through independent initiatives by farmers and consumers aimed at a more rational use of natural resources, but these efforts will always be of limited scope as long as the state is not involved.[2]

If Latin American government leaders were truly committed to promoting development models less damaging to natural resources, it would probably be feasible to do so. Major changes would be necessary in policies, institutions and technological patterns, but it could be done. However, this would involve high costs for certain groups, who would have to give up activities that pollute or degrade natural resources, finance efforts to incorporate 'cleaner' technologies, invest in measures to reduce poverty and accept a greater degree of grassroots participation in decision-making.[3]

The situation is complicated even more because the groups that pay the above mentioned costs to protect the environment are not necessarily the same groups that will reap the benefits. In the current context, a peasant who decides not to deforest gains nothing from the fact that the genetic resources of the forest may be potentially valuable to the pharmaceutical industry. Likewise, the owners of polluting factories rarely live in the cities or neighbourhoods they contaminate. If a transnational corporation degrades a certain area, it can often move to another which is not yet degraded.

Some environmental problems, although not all, have the additional difficulty that they require large investments that only mature over the medium or long term. For example, many of the benefits of investing in soil conservation to prolong the useful life of hydroelectric dams will not be visible until the middle of the next century. Ten or fifteen years may go by before investments in education provide economic payoffs. The impact of the current deforestation of the Amazon on the world's climate will only be felt after several decades. It is generally less expensive to prevent environmental problems than to solve them once they occur (and some types of environmental degradation are irreversible), but this implies foresight and a planning capability that does not always exist.

For all these reasons, it is far from trivial to ask the question of who, if anyone, has a sufficient stake in environmental concerns to overcome all of the well known barriers to collective action and to mobilize sufficient resources for achieving environmentally friendly policies. In recent years, economists in Latin America have begun to pay more attention to the environment. But their studies have tended to take a cost-benefit approach for the purpose of making recommendations. They just assume that the persons receiving this information will make the appropriate decisions, or that if they don't, it is due to 'political' factors which are considered outside the scope of their 'technical' analysis.

A key factor in determining whether or not environmental reforms are carried out is the analysis, cohesion, capacity, and strength of social forces pushing for these reforms. We should recognise that the environment is not one single issue, but rather a multitude of different specific topics, each of which has its own characteristics and decision-making dynamic. A group may favour certain environmental policies and oppose others; some initiatives have progressed, while others have not. There is not one environmental movement, but many, and it will probably always be that way. Nevertheless, it is still possible to explore some of the general tendencies, in the hope of motivating more specific studies in the future.

My basic hypothesis is that government leaders will not take strong measures to reduce natural resource deterioration unless they are pressured to do so. This hypothesis is based on the empirical observation that no major historic political or social reform has occurred in Latin America without social struggle. Governing groups have rarely promoted reforms to benefit the popular sectors or to protect natural resources, without being subjected first to heavy pressure.[4] Although the state has a certain degree of autonomy, policy-makers are generally concerned with the political viability of their decisions. If no one is pushing for the management of natural resources based on social criteria, it may mean that it is not politically viable.

Since social pressure is required to bring about deep reform, it is essential that we understand both the potential and the limitations of the social forces involved.

Extra-regional players

The emergence of strong environmental movements in the developed countries is partly due to the existence of important groups in those countries with high enough incomes to concern themselves not only with day-to-day survival but also with the quality of life in a broader sense. Without ignoring the historical presence of various environmental initiatives in Latin America itself, one could argue that recent concern about natural resources and the environment in the region is linked to the political, economic, and intellectual influence from the United States and Europe.

Following the 1972 United Nations Conference on the Human Environment in Stockholm, environmental movements in the North began to pressure their governments and international lending and technical cooperation organizations to support more benign environmental policies in Latin America.[5] They also began to operate directly in the region, opening offices and providing support to projects in many countries.

Pressure by developed countries for diverse environmental measures in Latin America has taken various forms. Many Northern NGOs have concentrated on narrow environmental issues such as the extinction of plant and animal species, rainforest preservation and pesticide and toxic waste contamination. Over time, however, some of them have broadened their concerns to include wider issues such as how structural adjustment policies, trade liberalization, and the decision-making mechanisms within international financial and technical assistance institutions affect environmental problems. Since the United Nations Conference on Environment and Development (UNCED) in Rio de Janeiro in 1992, a growing number have even begun to conceptualize the issue in terms of 'sustainable development', incorporating social as well as environmental issues and emphasizing the relation between the two.

Developed country environmental organizations have influenced policy in Latin America through various mechanisms. NGOs such as Greenpeace, the Rainforest Action Network, and the Pesticide Action Network have organized international campaigns to publicly denounce environmental dangers and pressure Latin American governments to take specific policy measures such as prohibiting the importation of toxic wastes and banned pesticides and taking steps to reduce deforestation and the extinction of plant and animal species. In 1983 a coalition of major US environmental organizations launched a successful campaign to pressure the multilateral development banks to pay more attention to environmental issues and stop financing certain projects accused of harming the environment.[6] Most major environmental organizations have full-time lobbyists who seek to influence Congress and the European Parliaments regarding funding for natural resource projects, international environmental conventions, policies towards international financial institutions, trade agreements, and other international issues. A recent example of the power of these groups was their success in convincing the Clinton Administration to negotiate an environmental side agreement to the North American Free Trade Agreement (NAFTA).

Of those international NGOs either working directly in Latin American or financing the activities of local environmental organizations, Conservation International, the Nature Conservancy, the International Union for the Conservation of Nature and Natural Resources (IUCN), Rodale, and the World Wildlife Fund figure prominently. Most of these are focusing on national park management. Several of them have been involved in 'debt for nature' swaps in Bolivia, Costa Rica, Ecuador, and the Dominican Republic, which have allowed them to obtain local currency funds for conservation activities by agreeing to purchase highly-discounted government debts.[7] Many traditional international private voluntary organizations (PVOs) have begun promoting natural resource projects related to soil conservation, pesticide management, and environmental education. Greenpeace, World Resources Institute (WRI), and other organizations provide technical assistance and advice to Latin American environmental organizations, and there are several active communications networks tying North and South NGOs together through electronic mail. This increased collaboration has been reflected in the International Forum of NGOs and Social Movements at UNCED, in international coordination around the General Agreement on Tariffs and Trading (GATT) negotiations, and in the negotiations related to the Biodiversity Convention and other international agreements.

The influences from the developed world have certainly helped to make environmental issues more visible in Latin America. They have also helped finance a growing number of Latin American environmentalists and natural resource projects. They have brought about specific policy reforms in areas such as livestock subsidies in the Amazon and pesticide regulations, and contributed to the creation of environmental ministries and agencies throughout the region. On the other hand, they have had little or no impact (and perhaps in some cases even a negative impact) on the underlying causes of environmental degradation such as poverty, the short-term time horizons of Latin American societies, the region's heavy dependence on environmentally degrading activities for export revenues, and the limited capacity of Latin American governments to regulate the use of collective resources.[8]

For the most part, Northern governments and NGOs have tended to view the environment as if it were a separate sector, rather than an integral part of how production and society are organized. As a result there has been much more attention on cleaning up and protecting rather than on providing incentives for a more sustainable pattern of development. This curative, rather than preventative, approach tends to be expensive and, when combined with the high overhead costs of international assistance, provides only localized results. To date, policy dialogues have focused more on legislation and the creation of environmental bureaucracies than on adjustment of macro-economic and trade policies and changes in resource flows, where they might have more impact. The number of cases where major environmental policy reforms have been an important condition for loan approval or trade concessions are still few and far between.

The decline in American and European bilateral development assistance to Latin America, prompted by the developed countries' retreat into domestic issues and by their focus on Eastern Europe, has reduced the North's leverage in environmental matters. A recent example of this can be seen in the proposal by Central American governments to create an 'Alliance for Sustainable Development' between the United States and Central America, in the hope of obtaining greater foreign assistance. The United States responded to this initiative by welcoming the concept but made it clear that no new funds were likely to be forthcoming. This clearly limited US influence in the initiative.

The urban middle classes

The environmental movement in the developed countries forms part of what is now called the 'new social movements.'[9] The social base of these movements are middle-class professionals and youth with a certain degree of education, who join these movements not so much to receive direct material benefits but rather to express their frustration with government institutions and modern society and their desire for an identity and for social spaces where they can express themselves and influence society.[10]

The phenomenon of the new social movements is also present in Latin America, primarily in large urban centres such as Santiago, Chile, Mexico City, Caracas, and São Paulo, Brazil, where numerous environmental organizations of middle-class extraction have emerged. For example, in Brazil, the country with the strongest environmentalist movement, of some seven hundred environmental organizations in 1989, ninety per cent were located in the prosperous south-east, with over a hundred in the city of São Paulo alone.[11] In 1986 São Paulo elected its first congressman on an environmental platform and around the same time the Green Party achieved eight per cent of the votes in the elections for governor of Rio de Janeiro. In both cases the basic constituency was middle-class professionals.[12] Brazilian environmental groups have concentrated on protecting specific ecosystems such as the Atlantic Forests, the Amazon, and the watersheds of Paraná, and in offering environmental education. They also carried out a successful lobbying campaign to include environmental considerations in the 1988 Brazilian constitution and played an important role in the preparations for UNCED. In Chile, middle-class environmentalists have played a major role in bringing national attention to the plight of native forests and were influential in the debates surrounding the recently adopted Law of the Environment. In Venezuela, they succeeded in cancelling the Trans-Amazon Rally in 1987, a proposed car race from Venezuela to Brazil.[13]

Beyond the organized environmental movement, there is also a much broader universe of middle-class public opinion that is sensitive to the environmentalist messages transmitted by the mass media. Much of the middle class suffers directly from air, water and food pollution and traffic congestion, which serve as a constant reminder of the problems affecting the environment. These groups react with concern to the news about environmental catastrophes such as the release of nuclear radiation in Chernobyl, floods or droughts attributed to environmental imbalances, and oil spills by shipwrecked boats.

The middle classes' interest in the environment tends to centre around urban pollution and some widely publicized symbols of the destruction of nature, such as deforestation in the Amazon, the danger of extinction of certain well-known animal and tree species, and the struggles of indigenous peoples and rubber tappers. The influence of these groups has been sufficiently strong to oblige practically all of Latin America's political parties to incorporate environmental planks into their political platforms and to get the governments of Chile and Mexico to take concrete measures, though weak ones, to reduce air pollution in the capital cities.

On the other hand, the urban middle classes' concern with the environment is sporadic and based more on images than on a clear understanding of the problems and their underlying causes. They have relatively little interest in wider issues related to the patterns of natural resource use or less 'sexy' environmental problems. Middle-class public opinion responds emotionally to the environmental symbols of popular culture and the mass media but this response is only occasionally converted into effective pressure. This hinders the consolidation of stable membership organizations that can maintain pressure for reform over time. The economic crises in many Latin American countries and the cutbacks in government expenditures have tended to reduce the incomes of many middle-class sectors, forcing them to pay more attention to their own economic survival, and leaving less space for environmental concerns. Moreover, it is important to remember that only a few Latin American countries have a large, concentrated, urban middle-class population.

Groups motivated directly by material interests

At least four groups have direct and immediate material interests in environmental management: companies and farmers who sell to the so-called 'green markets,' producers and communities affected by critical pollution problems, indigenous movements and professional environmentalists.

There is a growing number of companies and small-farmer groups interested in taking advantage of new opportunities offered by the markets for ecotourism, organic foods, recyclable containers, and products from sustainably managed forests.[14] These farmers have a direct interest in the growth of environmental awareness, since it contributes to expanding their markets.

There are also a large number of businesses who sell traditional products, but who have incorporated ecological messages or symbols into their advertising because market research indicates this will help sell their products. These companies' advertising, although sometimes misleading, also helps to keep the public's attention on the environment.

The 'green industry' which has probably had the most influence on national politics is tourism in Costa Rica. Having grown quite rapidly in recent years, tourism is now Costa Rica's largest foreign exchange earner. In 1993 it brought in over $500 million in gross revenues. Although the activity involved is often characterized as ecotourism, it actually combines ecological activities such as visits to national parks with traditional tourist attractions such as beaches and museums. The tourist industry has clearly profited from Costa Rica's international reputation as a country which protects its environment, and despite the fact that it has come into conflict with the government over specific environmental regulations, the industry is a major supporter of government policies which help project an environmentally friendly image.

At the other extreme are producers and communities suffering from enormous economic losses as a result of pollution and natural resource degradation. These include artisanal fishermen affected by water pollution, petty extractors (such as rubber tappers or Brazil nut gatherers) threatened by forest destruction, communities living near toxic waste dumps and other types of waste, and farmers harmed by pollution from the mining or petroleum industry, among others.

These groups have started to protest and bring their grievances to the national spotlight. The most famous case, of course, was Chico Mendes' struggle to protect the rubber tappers of Acre, Brazil from encroachment by ranchers and loggers. But there have been others too, such as the movement of the Penwalt Battery workers in Nicaragua to gain compensation for having been exposed over many years to excessive levels of mercury, or the protests by Mexican peasants against oil spills and pollution of the rivers in Tabasco and arsenic poisoning 'due to the overuse of the aquifer in Comarca Lagunera and irrigation with sewer water in the Mezquital Valley'.[15] Labour unions and environmental organizations in northern Mexico led the fight against toxic wastes produced by border industries, and coordinated closely with environmental organizations in the south west of the United States seeking to incorporate environmental provisions into the North American Free Trade Agreement.

In the case of indigenous movements, the fact that the indigenous peoples have lived in a certain degree of harmony with their natural environment has given them considerable legitimacy in the eyes of public opinion, and has become a powerful argument in support of their territorial demands.[16] For them, a successful outcome of their territorial struggles is critical to their survival as peoples. The recent struggles of the Yanomami Indians in the Brazilian Amazon and of the Mayan Indians of Chiapas are well known. But indigenous movements for land rights in humid tropical lowlands with environmental undertones have also played important roles in national politics in Bolivia, Ecuador, Panama, and Venezuela.

Finally, there is an important group of people directly employed in environmental activities, who have an evident interest in maintaining support for environmental issues. This includes public officials in entities and projects dealing with environmental concerns, academics, environmental journalists, and employees of NGOs working in this field. Most Latin American countries today have at least one public environmental ministry or agency plus several environmental NGOs with professional staff, a dozen or so foreign-financed environmental projects, and a handful of university departments with professors working on environmental issues. This group is not large but it exercises significant influence because its members are in a position to mould public opinion. In particular, the participation of scientists in environmental movements has lent greater credibility to the movements' warnings.

None of those with a material interest in environmental protection are currently a major force in Latin American societies, but their influence is growing rapidly. With the exception of the academics and professional environmentalists (and the tourism industry in Costa Rica), their presence tends to be local and their profile at a national level very minimal. By themselves, they probably do not have sufficient force to spur deep reforms, but they have helped keep the subject of the environment in the public eye and have achieved concrete results in certain areas.

The social justice movement

Jorge Castañeda claims that 'in 1980, 120 million Latin Americans, or 39 per cent of the area's population, lived in poverty; by 1985 the number had grown to 160-170 million; toward the end of the decade it was estimated at the appalling figure of 240 million.'[17] The main concern of this majority of poor Latin Americans is to achieve a dignified standard of living for themselves and their children. Historically, this concern had its political reflection in the development of left-wing political parties (communist, populist, social democratic), and of trade unions, farmers' organizations, and community movements.

The strengthening of neoliberal currents in Latin America, the loss of confidence in the traditional parties, and the disappearance of the socialist bloc led by the former Soviet Union have weakened the Latin American Left. Nevertheless, there are still some parties, organizations and movements of certain importance whose main concern is to fight for social justice. Left or left-centre parties such as the Workers Party (PT) in Brazil, Revolutionary Democratic Party (PRD) in Mexico, the Socialist Party in Chile, the Radical Cause and Movement Towards Socialism (MAS) in Venezuela, the Sandinistas in Nicaragua (FSLN), and the Farabundo Martí Front for National Liberation (FMLN) in El Salvador form part of governing coalitions or have reasonable hopes of doing so in the next few years. There has been a resurgence of farmers' organizations made up of small- and medium-scale farmers in Central America, several Andean countries, and Paraguay. There has also been a resurgence of human rights groups, women's organizations, community organizations, and service-oriented non-governmental organizations, concerned with social justice issues.

Given the extreme and massive poverty in Latin America, the possibility of linking environmental problems with the struggle for social justice is practically the only way to make environmental issues relevant to the bulk of the population. Thus, inclusion of environmental issues in the agenda of social justice movements is essential to developing mass constituencies for major policy reforms related to the environment. Without this convergence, some environmental reforms may still take place, but they would be elite reforms, without the benefits of wide and democratic popular participation.

In the last few years, certain sectors of the Latin American left have begun to pay closer attention to the environment. There is a strong environmental current within the Brazilian Workers Party and, to a lesser extent, the Chilean Socialist Party. There is also a large number of grassroots organizations and NGOs that consider themselves 'on the left' who have taken an interest in the environment. These groups are aware that concern for the environment has the potential to become an issue with almost universal support, as was the banner of peace in the past, and that this will allow them to appeal to public opinion as a whole, not just the sectors directly involved in the struggles of specific social groups. They are also aware that the environment, together with the social concerns arising from it, constitutes a weak point in the neoliberal paradigm. A critique of neoliberals' environmental policy offers fertile ground for a more general questioning of neoliberal thought.

Some small farmers' organizations, such as the 'Plan de Ayala National Coordinating Council' in Mexico, the Ecuadorean farmers unions, and the 'Council of Small and Medium Farmers – Justice and Development' in Costa Rica, have seen the emergence of environmental concerns as an opportunity to construct new alliances, by arguing that small-scale production can play a role in preserving diversified production systems and in conserving natural resources. They are aware that there is no longer much urban support for land reform, subsidized farm credit or price supports for farmers. But by stressing their role as stewards of their environment, small farmers have found a way to gain support from urban middle classes and foreign aid agencies.

At the same time, there is reason to doubt the depth of the Latin American left's commitment to the environment. For some, the environment is no more than a superficial fad imported from the northern countries. They argue, with some reason, that many conservationists are more concerned with animals and plants than with people. Although almost every party gives at least rhetorical support to environmental issues, parties such as the PRD in Mexico, the FMLN in El Salvador, and the Sandinista Front in Nicaragua have shown little real interest in 'green issues'. There seems to be much greater support for environmental causes among the middle-class supporters of the left-wing parties than among their poorer constituencies or among party bureaucrats. Certain trade unions are afraid that environmental restrictions will limit job opportunities. Furthermore, it is relatively easy to be an environmentalist from the opposition. It is not so clear that left-wing movements will give high priority to environmental issues once they acquire governmental responsibilities.[18]

As long as environmental concerns are associated in the public view with wildlife protection and biodiversity, protection of the ozone layer, and global climatic change, it seems unlikely that they will be able to win the attention of the millions of Latin Americans who are on the edge of survival. There are other issues, however, such as access to safe water supplies and firewood, soil conservation on marginal lands, safety from pesticide poisoning and protection from other toxic chemicals in the work place, and reproductive rights that could potentially appeal to such groups.

The links between poverty and environmental degradation could also serve as the basis for alliances between those concerned with social justice and the conservationists. This is the (as yet mostly unrealized) hope of UNCED. Both groups should share an interest in finding alternatives for the marginal hillside farmers of Central America who are producing the sediments which threaten the hydroelectric plants, the slash-and-burn farmers of the tropical lowlands who are destroying the forests which could be used to produce pharmaceutical and agricultural germplasm, and the urban masses who contaminate the cities with their human wastes because they have no other way to dispose of them.

One encouraging sign in this regard is the decision by the Figueres government in Costa Rica to make sustainable development the theme of its administration. Figueres campaigned largely on the issues of poverty and the environment, and has sought to develop a new social democratic approach capable of responding to these problems within the broader context of trade liberalization and the discrediting of previous attempts at social reform. Through these efforts, Figueres clearly hopes to win support from the international financial institutions and bilateral agencies and the local tourist industry, but he also appears interested in building a broader domestic constituency for environmental reform. If he succeeds, the Costa Rican experience may hold important lessons for other Latin American countries.

Conclusions

The feasibility of generating a political will to manage natural resources in a rational manner in Latin America is still uncertain. Although pressures in favour of environmental reforms have greatly increased, they continue to be weak currents compared with the overriding concerns with generating economic growth, reducing poverty, and achieving political stability.

Current pressures for better environmental policies come from a wide variety of disparate sources. Even when they are successful at getting legislation passed or achieving formal government recognition, they often fail to bring about real change, due to the governments' weak implementation capacity or a mistaken diagnosis of the underlying problems and possible solutions. The tendency to consider the environment as a separate sector has aggravated this problem because it means that all of the underlying processes which create environmental problems continue and it is up to the environmental sector to try to reduce their harmful impact once they have already created significant damage.

Some of the most vocal movements for environmental reform exist only at the local level. For them, the recent trend towards more decentralized government in Latin America may open up new opportunities for influence, which did not previously exist at the national level.

For the vast majority of Latin Americans, the basic issue is not environmental conservation but day-to-day survival. Getting these popular sectors involved and making the environmental movement something more than a dim reflection of developed country environmental concerns will only happen once the immediate impact of environmental degradation on peoples' daily lives is addressed. Only then will environmental issues secure active support from the parties and organizations which have traditionally served as spokespeople and mobilizers of the urban and rural poor. There are some indications that this is beginning to happen but it remains a slow process.

[1] The opinions stated herein are those of the author and do not necessarily reflect the position of IICA. The document was enriched through discussion with Eduardo Baumeister, Fabiola Campillo, Manuel Chiriboga, Ricardo Costa, Charlotte Elton, Gonzalo Estefanell, Roberto Haudry, Henri Hocde, Roberto Martínez Nogueira, David Mayhre, Orlando Plaza, Grettel McVane, Lori Ann Thrupp, Eduardo Trigo, and participants in the seminar on the economics of sustainable development in Central America, held in Catalina, Costa Rica, June 20-23, 1993.

[2] In this document, the rational management of natural resources is understood to mean rational from the perspective of society, since over-exploitation of natural resources and contamination can be rational from an individual point of view.

[3] Reducing poverty and boosting democratisation are essential for reducing environmental deterioration. While the poor are not the main culprits behind environmental destruction, if measures are not taken to reduce poverty, it will be difficult to revert erosion on hillsides and deforestation for firewood linked to the advance of the agricultural frontier, or problems associated with sanitation and personal hygiene, or high birth rates.

[4] This applies even to government leaders with some autonomy from the ruling classes. For example, agrarian reforms carried out by Cárdenas in Mexico and by military government leaders in Honduras and Peru, Torrijos' struggle in Panama to maintain sovereignty over the

Canal Zone, the Peronist reforms in Argentina and Calderón Guardia's social reforms in Costa Rica were all preceded by strong social pressure.

[5] Barbara Bramble and Gareth Porter, 'Non-Governmental Organizations and the Making of US International Environmental Policy', pp. 313-353, in *The International Politics of the Environment*, Andrew Hurrell; Benedict Kingsbury eds. (Oxford: Clarendon Press) 1992.

[6] Ibid.

[7] See Amparo Chantada 'Los canjes de deuda por naturaleza, el caso dominicano', *Nueva Sociedad*, No. 122 (November – December) 1992: 164-175.

[8] See Hernán Durán de la Fuente, 'Contaminación industrial y urbana: opciones de política', *Revista de la CEPAL*, No. 44 (August), 1991: 137-148; Fernando Fajnzylber, 'Competitividad internacional: evolución y lecciones', *Revista de la CEPAL*, No. 36 (December), 1988: 7-24; Héctor Leis, 'El rol educativo del ambientalismo en la política mundial', *Nueva Sociedad*, No. 122 (November – December) 1992: 116-127; David Reed ed. *Structural Adjustment and the Environment* Westview Press, Boulder, Colorado, 1992.

[9] Other movements associated with this category are the feminist, student and pacifist movements. See Frederick Buttel, 'Environmentalism: Origins, Processes, and Implications for Rural Social Change', *Rural Sociology*, Vol. 57, No. 1, 1992: 1-27.

[10] Buttell, op cit.

[11] See Eduardo Viola, 'El ambientalismo brasileño, de la denuncia y concientización a la institucionalización y el desarrollo sustentable', *Nueva Sociedad*, No. 122 (November – December) 1992: 138-155.

[12] Enrique Leff, 'Environmentalism: Fusing Red and Green', *NACLA Report on the Americas*, Vol. XXV, No. 5 (May) 1992: 35-38.

[13] Leff, 1992, op cit.

[14] Stephan Schmidheiny, *Cambiando el rumbo, una perspectiva global del empresariado para el desarrollo y medio ambiente*, Fondo de Cultura Económica, México D.F. 1992.

[15] Ibid.

[16] Adriana Hurtado and Enrique Sánchez, 'Introducción – documento de reflexión síntesis. Situación de propiedad, aprovechamiento y manejo de los recursos naturales en los territorios indígenas en áreas bajas de selva tropical', 11-36, in *Derechos territoriales indígenas y ecología en las selvas tropicales de América* (Bogotá: Fundación Gaia / CEREC), 1992.

[17] Jorge G. Castañeda *Utopia Unarmed, The Latin American Left After the Cold War*, Alfred A. Knopf, New York 1993, p. 498.

[18] See Buttell op cit for an interesting discussion of the relationship between traditional social democratic movements and the environmental movement in developed countries.

2
INDIGENOUS COMMUNITIES AND THEIR ENVIRONMENTS

Native peoples and sustainable development*

Al Gedicks

In 1987 the United Nations' World Commission on Environment and Development, also known as the Brundtland Commission (after its chair, Prime Minister Gro Harlem Brundtland of Norway), published *Our Common Future*.[1] The key concept of the report was 'sustainable development,' defined as 'development that meets the needs of the present without compromising the ability of future generations to meet their own needs.' The report also acknowledged the crucial role of culture as an adaptive mechanism. In surveying the impact of externally-imposed development upon native cultures in the remote regions of the globe, the Brundtland report noted the 'terrible irony' that 'it tends to destroy the only cultures that have proved able to thrive in these environments'. To counter this tendency, the report advocated 'the recognition and protection of [native cultures'] traditional rights to land and the other resources that sustain their way of life – rights they may define in terms that do not fit into standard legal systems'.[2]

The concept of sustainable development received a great deal of attention in the discussions leading up to the UN Conference on the Environment and Development, or the Earth Summit, held in Rio de Janeiro, Brazil in June 1992. Thomas N Gladwin, a professor of management and international business at New York University, described the concept as 'the cutting edge of social and economic reform'.[3] But the cutting edge of this concept has been dulled as practically every major institution in the world economy – from multinational mining and logging companies to the World Bank – has embraced the concept. Even the International Atomic Energy Agency claimed that 'the supply of energy for economic growth in a sustainable and environmentally acceptable manner is a central activity in the Agency's programme'.[4] The real question that needs to be answered is: 'sustainability of what and for whom?'[5]

The different interpretations of sustainable development have assumed a particular urgency in the context of recent international proposals for preserving the biological diversity in the world's rainforests. At least half of all known species are contained in the tropical rainforests.[6] One of the plans for preserving world biodiversity is the Tropical Forestry Action Plan, drawn up by the World Resources Institute in cooperation with the World Bank, the United Nations Development Program, and the Food and Agricultural Organization. The plan envisions an $8 billion action programme to encourage production of fuel wood, commercial forestry, reforestation, and conservation of tropical forest ecosystems. Among the major short-comings of the plan, according to a recent report by Marcus Colchester of the World Rainforest Movement, is that 'it paid little attention to the needs and rights of forest dwellers and seemed unduly focused on funding commercial forestry and wood-based industries, while failing to identify the real causes of deforestation'.[7] Among the real causes of deforestation, as we have seen, are large-scale agriculture, cattle breeding, road construction, logging, oil drilling, mining, and dams – all the activities financed by major financial institutions such as the World Bank. The implementation of the Tropical Forestry Action Plan at the national level 'will promote a massive expansion of logging in primary forests. Despite the fact that rainforest logging is not being carried out in a sustainable fashion and is itself one of the principal causes of deforestation, under the Tropical Forestry Action Plan logging in primary forests will intensify.'[8]

The Tropical Forestry Action Plan is just one example of the ethnocentric and neocolonial perspectives that ignore the role of native peoples in maintaining sustainable resource systems in the areas they still inhabit. Another example is the increasingly popular 'debt for nature swaps' being promoted by banks and environmental groups. Here again, the knowledge and experience of native peoples is frequently overlooked as parts of the rainforest are set aside for conservation in return for the cancellation of foreign debt. 'You must understand,' says Santos Adam Afusa, a Peruvian Indian and member of the coordinating body for the Indigenous Peoples' Organization of the Amazon Basin, 'that we try to maintain the forest as our ancestors have maintained, protected, and taken care of it. That is extremely important because the destruction of the forest would not only ruin our homelands but would have a worldwide effect. The issue is not just to protect the forest; the issue is also to [rely on] the people that are able and have for millennia been able to protect and maintain the forest.'[9] Indeed, as Winona LaDuke has pointed out, there is a widespread assumption among those promoting these megaprojects that 'indigenous people do not have their own economic systems so someone should come and give them one. This development program is a war on subsistence.'[10] The alternative to this 'war on subsistence' is a type of sustainable development which recognizes 'the right to exercise its authority and jurisdiction over the corresponding territory'.[11]

We know from studies in ethnobotany that native peoples are often aware of differences in nature which are invisible to specialists from the outside. For example, in one Aymara community in the Bolivian Altiplano, peasant households named 38 'sweet' and nine 'bitter' varieties of potato that they themselves cultivate.[12] 'Their interest in maintaining crop diversity is based not on a belief in diversity for its own sake, but on the knowledge that diversity reduces their environmental vulnerability.'[13] Darrell Posey, an ethnobiologist, has reached similar conclusions, based upon his research with the Kayapo in Brazil. According to Posey, the Kayapo use the heterogeneity of the Amazon to broaden their resource potential.[14] And after surveying native land-use practices in Central America, one study concludes that 'there are no other land-use models for the tropical rainforest that preserve ecological stability or biological diversity as efficiently as those of the indigenous groups presently encountered there'.[15]

We cannot expect to preserve fragile ecosystems while the native peoples who live in these areas are dispossessed and forcibly dislocated. This is the foundation of the emerging unity between native peoples and the international conservation movement. As ecologically-destructive megaprojects continue to penetrate the world's resource frontiers, the global problems of deforestation, desertification, depletion of fisheries, and soil erosion are major concerns of both groups.[16] In 1970, UNESCO provided a conceptual framework for integrating the goals of conservationists and native peoples with the creation of the 'Man and the biosphere' programme. Among the major objectives of the programme are the conservation of ecosystems which are ecologically self-sustaining, the proportion of research and monitoring on their appropriate use and management, and the involvement of native peoples in all phases of the project. Native rights advocate Bernard Nietschmann suggests that Man and the Biosphere reserves 'are among the most attractive for expanding the role of traditional peoples in reserve design, management, and interactive research objectives'.[17]

One of the more successful applications of the biosphere reserve concept is the Kuna Wildlands Project in Panama. The project area covers 60,000 hectares of protected wildlands on the Kuna Yala native reserve on the northeastern coast of Panama. Included in this reserve are coral reefs, islands, mangroves, coastal lagoons, gallery forest, and evergreen hardwood forest.[18] One of the first priorities of the park was the physical demarcation and protection of the entire reserve boundary of the Kuna people. Much of the success of the Kuna reserve can be attributed to their high degree of social organization and cohesion and a strong tradition of autonomy and self-reliance.[19] After surveying the experiences of native peoples with national parks in northern Canada, Alaska, and northern Australia, one study found that the most successful cases of native-conservationist relations occurred where there were organizational vehicles (native corporations, aboriginal land councils) to ensure native control over the land and input into planning.[20]

As native experience and knowledge is taken into account in managing biosphere reserves, there is an implicit challenge to the prevailing Western model of technology transfer from the 'advanced' societies to the 'less advanced' societies. The more we learn about native economies, the more appreciation we have for 'small-scale economic development' models based on technologies that are low-cost, labour-intensive, and ecologically sound. In an explicit challenge to the entire 'industrial model' of economic development, the World Council of Indigenous Peoples called upon the international community to recognize the important contributions of these native technologies to sustainable development. 'The trend of industrial technology,' according to the council statement, 'has reduced the variety of technology that may be available for human development. The ability of human society to adapt to a variety of environmental and social changes has been correspondingly reduced. The contributions of diverse populations to the pool of technologies must be expanded rather than narrowed. Human diversity must be matched by technological diversity. The constant recognition of the value of technologies appropriate to political, human, and ecological circumstances is essential to achieving balanced development throughout the world.'[21]

Clearly, the goal of sustainable development is inseparable from the goal of maintaining cultural diversity. In a recent Worldwatch paper, Alan Thein Durning argues that 'the world's dominant cultures cannot sustain the earth's ecological health without the aid of the world's endangered cultures. Biological diversity – of paramount importance both to sustaining viable ecosystems and to improving human existence through scientific advances – is inextricably linked to cultural diversity.'[22] The sooner we stop labelling 'native issues' as something separate and distinct from our own survival, the sooner we will appreciate the critical interconnections of the world's ecosystems and social systems. 'At some point' says Thomas Meredith, a geographer at McGill University in Montreal, Canada, 'it may be possible to recognize cultural adaptive diversity as having the same conservation value as genetic adaptive diversity... At the very least, however, the many culture groups whose knowledge and values lead them to favour environmental and economic circumstances that seem alien to modern Western culture must be recognized, as they have never been in practice, and given a share in determining the fate of their own habitat.'[23] To facilitate this process, Meredith urges an appropriate process of environmental impact assessment, involving native peoples and including the study of 'socio-ecosystems.'

One of the most far-reaching recommendations of the World Conference of Indigenous Peoples on Territory, Environment, and Development, held at Kari-Oca Villages in Brazil in May 1992, was the call for a war crimes tribunal, modelled on the Nuremberg trials after World War II, to focus international public attention on those corporations and nation-states which impose megaprojects on native lands without obtaining consent or involving native peoples in the decision-making process. Such failures should be considered 'crimes against indigenous peoples' and those responsible should be tried in a 'world tribunal within the control of indigenous peoples set for such a purpose'.[24] The September 1992 World Uranium Hearing in Salzburg, Austria was a step in that direction. Such a tribunal could serve as a focal point for bringing together regional, national, and international native-environmentalist alliances to provide an effective counterweight to the ability of multinational corporations and nation-states to wage future resource wars against native peoples.

Conclusions

In assessing the priorities for advocacy work on behalf of native peoples, Cultural Survival has observed that 'the era of the resource wars is just beginning. At stake is not only the issue of ownership, but the value of resources and who has the right to manage and consume them.'[25] Increasingly, native peoples are at the forefront of these battles because their lands are directly threatened by mining, lumber, hydroelectric, or military projects. Until quite recently, native peoples have had to defend themselves against multinational corporations and nation-states using their own very limited resources and with hardly any notice from the rest of the world. The situation has radically changed over the past decade. The integral connections between native survival and environmental protection have become apparent to even the most conservative environmental organizations. Now the assertion of native land rights takes place in the context of an environmental movement that is prepared to appreciate the knowledge native people have about their own environment and to accept native leadership in environmental battles.

*Reprinted with permission from Al Gedicks *The New Resource Wars: Native and Environmental Struggles Against Multinational Corporations*, South End Press 1993.

[1] *World Commission on Environment and Development Report,*, 1987, p.43.
[2] Ibid, p. 115,
[3] Emily T. Smith, 'Growth vs. Environment' *Business Week* No 3265, 11 May.
[4] Thijs de la Court, *Beyond Brundtland: Green Development in the 1990s*. Translated by Ed Bayens and Nigel Harle, Zed Books, London.
[5] Thomas C. Meredith, Environmental Impact Assessment, Cultural Diversity, and Sustainable Rural Development' *Environmental Impact Assessment Review* Vol 12 Nos 1/2, March/June 1992.
[6] Andrew Gray, 'Between the Spice of Life and the Melting Pot: Biodiversity Conservation and its Impact on Indigenous Peoples', *International Work Group for Indigenous Affairs*, Copenhagen, August 1991
[7] Marcus Colchester, 'The Tropical Forestry Action Plan: What Progress' quoted in Andrew Gray op cit.
[8] Ibid p.2
[9] Adam Santos Afusa, 'Indigenous People Can Save Rain Forests' in Matthew Polesetsky (ed) *Global Resources: Opposing Viewpoints*, Greenhaven Press, San Diego, 1991.
[10] Winona LaDuke, 'Environmental Work: An Indigenous Perspective' *Northeast Indian Quarterly* Vol 8 No. 4 (Winter).
[11] 'Statement on Self-Determination by the Participants at the Indigenous Peoples Preparatory Meeting' *Yearbook 1987: Indigenous Peoples and Development*, IWGIA, Copenhagen.
[12] J. Dandler and C. Sage, 'What is happening to Andean potatoes? A view from the grassroots' *Development Dialogue* 1 1995, Uppsala, Sweden.
[13] Michael Redclift, *Sustainable Development: Exploring the Contradictions*, Methuen, London 1987.
[14] Darrell Posey, 'Nature and Indigenous Guidelines for New Amazonian Development Strategies: Understanding Biological Diversity through Ethnoecology' in J. Hemming (ed) *Change in the Amazon Basin* Vol 1: *Man's Impact on Forests and Rivers*, Manchester University Press, 1985.
[15] Brian Houseal et al, 'Indigenous Cultures and Protected Areas in Central America' *Cultural Survival Quarterly* Vol 9 No 1, February 1985.
[16] James C. Clad, 'Conservation and Indigenous Peoples: A Study of Convergent Interests' in John H. Bodley (ed) *Tribal Peoples and Development Issues: A Global Overview*, Mayfield Publishers, California, 1988.
[17] B. Nietschmann, 'Biosphere Reserves and Traditional Societies' in *Conservation, Science and Society: Contributions to the First International Biosphere Reserve Congress, Minsk, Byelorussia/USSR*, UNESCO, 1984.
[18] Houseal 1985, op cit, p.10.
[19] Ibid.
[20] J.E. Gardner and J.G. Nelson, 'National Parks and Native Peoples in Northern Canada, Alaska and Northern Australia' in John H. Bodley op cit.
[21] 'Submission Concerning the United Nations Third Development Decade, United Nations Economic and Social Council', World Council of Indigenous Peoples (WCIP), 1979.
[22] Alan Thien Durning, *Guardians of the Land: Indigenous Peoples and the Health of the Earth*, Paper No.112, Worldwatch Institute, Washington DC.
[23] Meredith 1992, op cit, p.126.
[24] Kari-Oca Declaration, 'Indigenous People Earth Charter', IWGIA Newsletter No.4, Nov-Dec 1992.
[25] 'Sharing the Wealth?' *Cultural Survival Quarterly* Vol 15 No 4, Fall 1991.

'The land of our ancestor's bones': Wichí peoples' struggle in the Argentine Chaco

Aidan Rankin

The Wichí Indians of northern Argentina and south-eastern Bolivia occupied and cultivated their land for thousands of years before Europeans 'discovered' it, drew borders and gave it alien names. Unlike the coastal Indians of what are now Argentina, Uruguay and Chile (almost all of whom fell victim to genocide, slavery and disease) the misery of the Wichí began this century. Since the 1900s, Argentina's last indigenous peoples have been dispossessed by settlers known as *criollos*, who regard Indian territory as an unpopulated frontier.[1] Wichí lands, which are fertile when carefully managed despite the Chaco region's dry climate, have been turned into an inhospitable dust bowl as cattle ranching and other unsustainable forms of commercial agriculture have been introduced.

Yet today, spurred on by the democratic opening that has affected the rest of Argentina in the last ten years, the Wichí are organizing themselves against the discrimination they experience in their daily lives. They have on several occasions successfully confronted the government of Salta province (where most of them live). In 1994, Wichí representative Francisco Pérez went to Geneva to address the United Nations Commission on Human Rights, an unprecedented event for an isolated people previously diffident in asserting their collective rights. With the help of Survival International, the Wichí have constructed a map of their entire territory and put into writing the rich oral history that reinforces their land claim. This project, whereby the oral communication of a people's history is transposed into written form, is a milestone in the cause of indigenous land rights throughout the world. It enables oral history, so often marginalized by Eurocentric interpretations of the past imposed by colonial authorities and national governments, to become a tool of political campaigning as well as a custom crucial to the survival of a beleaguered minority. For the Wichí, reclaiming history from the criollo interlopers is a first step towards the reclamation of their land and their right to determine their own future.

Occupation and violence: the colonization of Wichí land

> 'The man who first enclosed an area of land said "This is mine" and found others simple enough to believe him was the true founder of civil society.'

So wrote Jean-Jacques Rousseau in the mid-eighteenth century, an era when the dominant ideology of the West was shifting rapidly from paternalism to property and 'free trade'. Yet the Wichí of Argentina have never willingly accepted their subjugation. Like most indigenous peoples, they are bemused by Western ideas of private property and market economics. To them, there is an explicit link between the parlous state of their land under criollo control and the failure of the settlers to come to terms with the Indians' ancient culture. In the Wichí political struggle, the question of land rights (for indigenous peoples a prerequisite of human rights) finds common ground with the concerns of the environmental movement. The Argentine Chaco is a place where in the name of 'modernization' and 'development', an unsustainable way of living has been ruthlessly imposed and the lessons of thousands of years of successful land management discarded. The title *Green Guerrillas*, therefore, can usefully be applied to the Wichí (although the concept of guerrilla warfare is entirely alien to them). By demanding the return of the land they are subverting Western notions of progress as well as reclaiming what is rightfully theirs.

There are around 50,000 Wichí Indians surviving in the Chaco today. Before the arrival of the criollos, their society was organized into village communities. Each village has its own territory, but often six or seven villages will share the use of overlapping areas. Wichí communities usually consist of one or more clans. People belong to the mothers' clan and men move to their wife's village after they marry. Wichí houses are low and box-shaped, with mud-caked branches for walls and leafy sticks for roofs.

Argentina's Indian population had no conception of property beyond the most basic of personal possessions and the strong belief that their land belonged to them. They were therefore entirely unprepared for its seizure and enclosure by a new breed of colonist. To the Wichí, the fences erected by incoming criollos were symbols of an alien, hostile and baffling world. In 1991, for example, an elderly Wichí man named Qatsí (meaning 'He stays at home') described to Survival International an incident of persecution stemming from western-style property laws:

> It was an iguana that led me into trouble. One morning four years ago [my young son and I] set off... with three good tracking dogs. One of them followed the scent of an iguana, found it and chased it to its burrow. I had killed it and bagged it when my son said 'There is a young white man coming this way...' He was riding a bicycle and had four large dogs with him and a revolver.[2]
>
> 'What are you doing around here?' he asked. 'I own this land and I don't want Indians on it. I forbid you to hunt here. Then he

> fired at me, aiming at my head.... His first two shots missed.... His next shot rang past my head and the explosion tore open my eyebrow. Now I am half blind in that eye. The fourth shot hit my shoulder. The bullet is still there.

The Chaco is a semi-arid, thinly forested region straddling Argentina's borders with Paraguay and Bolivia to the north. It is divided between the provinces of Chaco, Salta and Formosa in the modern Argentine state. The Wichí live by hunting, fishing and farming and have maintained, in this apparently inhospitable region, a way of life that the veteran green campaigner, Rudolf Bahro, would describe as a 'communitarian subsistence economy'.[3] In the dry, cooler winter months, they depend on fish from the Pilcomayo River that runs through their territory. They catch over eighteen different species of fish, of which the commonest are afwuķna (catfish), atsá (a kind of salmon) and sichús (white shad). During the wet summers, they cultivate corn, watermelons, beans and pumpkins and throughout the year they hunt armadillo, deer, foxes, peccary and (as mentioned above) iguana. Their region is also noted for its many varieties of wild honey.

Throughout the twentieth century, Wichí society and the environment that nurtured it have come under systematic assault. Well-armed settlers, seeing the Indians' land as ideal ranching territory, have upset a fragile ecosystem by introducing herds of cattle. Where there were once rich grasslands, dotted with bushes and trees, there is now a virtual desert. In this alien terrain, most of the animals the Indians used to hunt have died out. As the Wichí themselves say: 'Our land is dead, and we are dying of hunger.'

The pain of occupation has been compounded by violence. To protect the areas they have fenced off from their rightful owners, the incoming criollos resort to acts of casual cruelty. In the early twentieth century, this reign of low-level repression was interspersed with bouts of extreme terror in which hundreds of Indians were killed. It is no small wonder that the Wichí call the colonists *ahatai* which, according to their oral history, 'is like our words for "spirit of the dead" and "the devil".'

In the West, there is a persistent belief that the environment and human rights are separate issues, requiring separate political prescriptions. This means that the right to breathe clean air is not accorded equal status with the right to vote or the right to own property. The Wichí use the language of environmentalism against the enclosure of their land. They rightly point out to the authorities that the environmental crisis of the Chaco began with the systematic violation of indigenous rights. At the same time, they challenge Argentina's civilian rulers to honour their commitment to democracy. But the Wichí campaign transcends conventional green activism, because its starting point is the survival of a people. The Wichí do not want the Buenos Aires or Salta legislatures to pass laws protecting their land, but to control and protect it themselves because it is theirs by right. Their first step towards self-determination has been to reclaim a history that links them intimately to their environment and their land.

Oral history and the land rights movement

Today, in the much vaunted 'new' Argentina of Alfonsín and Menem, the Wichí are determined to exploit the current fluidity of the political system to ensure that both regional and national authorities recognise their land claim. Their history, which (unwittingly echoing Puenzo) they have called 'the real history', records in faithful detail the colonization of Llakha Honhat (Our Land).

> The ahatai have always coveted Llakha Honhat, and they have used deceit and violence in order to take it from us...They didn't plant the trees; they don't keep the bees; the wild animals and fish don't belong to them... We have always lived here, since the time of creation – we are as much a part of Llakha Honhat as the trees that grow on it. Our land belongs to us because we belong to the land.[4]

Over the years, the Wichí have used their oral tradition to chronicle the gradual takeover of their land by outsiders armed with false promises backed up by force:

> Some said they had come to buy iguana skins and they would go at the end of the season. But soon they set up trading stores and stayed – without mentioning it to us, as though we didn't exist....One colonist even threatened us with war. 'Indians, what will you do without weapons if we make war on you?' he asked.[5]

As a non-literate culture, the Wichí have a long and very precise historical memory and a strong sense of place. Recently, they have become aware of the advantage this gives them in pressing for the return of their land. In 1987, the governor of Salta province passed a law which recognised the settlers' right to the land they had occupied and made only trifling concessions to the Indians. Two years later, with their patience and all regional campaigning channels exhausted, Wichí community leaders contacted Survival International. Together with two anthropologists, they began a project intended to prove conclusively that the criollos are living on stolen land. To do this, the Wichí people worked out a four-pronged strategy: to carry out a census of all the Indians in the region, to make a map of every village, to record and put in writing the long oral history of Llakha Honhat and to construct a large map of the entire region, showing all the places they had ever used and named.

The team spent six months working among the Wichí, who participated in every aspect of the work. They found that the oral history was crucial to the construction of the map, which contained over 1,000 place-names throughout the 400,000 hectares of the contested area. It gives a clear picture of the ancient nature of their residence, and of their profound knowledge and understanding of the land. The names identify the past and present villages of the Wichí, their hunting, fishing and gathering grounds, gardens, water-sources and places of religious or cultural significance. Some of the names are tragic or comic reminders of past events. Dog's Sighting is where thirsty Wichí were led to water by their hunting dogs, Great Bite is where an iguana bit a Wichí in the neck as he was digging it out of its lair, Sun Dies where some Wichí who were camping during a hunting expedition saw an eclipse of the sun and Paraguayan's Anus is the place where a drunken road-construction worker from across the border fell from his horse onto a tree stump. By contrast, Gun Is Fired is where the shots of the ahatai were first heard and Dead Children where a settler tried to hide the bodies of two young Indians he had murdered.

A problem with the 'real history' has been the lack of a precise chronology of the kind that westerners understand. Expressions like 'long ago' or 'in the very recent past' take the place of months and years, and no dates attach to the place names listed above. However, the team found that genealogical time was as reliable as conventional dates. Most Wichí historical accounts are prefaced with such indicators as 'in my grandfather's time' or 'my great aunt told me'. There is a strong continuity in the oral tradition that can cross generations, clans and communities. This way, the Wichí have maintained a collective memory that is much more reliable than the official version of Argentine history.

In 1991 the report and map were formally presented to the governor of Salta. Later that year, he signed a decree recognising the Wichí's ownership of 400,000 hectares of land and confirming that they should be awarded a single, communal title to the entire area. However, he left office hours later following mid-term elections. The Wichí still have not received title to their land, but the 1991 decree implicitly recognises the validity of the map and the accuracy of Wichí history. If implemented, it would mean that the Wichí would be recognised as the legal owners and custodians of their land. Although they do not seek to expel all the criollos from their land, the Indians intend to halt the destructive cattle ranching and to restore the Chaco to its previous ecological balance.

The experience of campaigning has given the Wichí Indians a new political confidence. They have now formed their own organization called, appropriately, Llakha Honhat, through which members of Indian communities can meet with officials at all levels of government. In 1994 a Wichí Indian made the trip to Geneva to put his people's case before the United Nations Commission on Human Rights. Like other indigenous peoples, such as the Innu of Canada, Australia's Aborigines and the Yanomami of Brazil, Argentina's long-forgotten Indians have now acquired a political voice. In spite of their prolonged sufferings and countless letters to the authorities which have gone unanswered, the soft-spoken Wichí have retained an optimistic faith in natural justice. They are rightly proud of their history, and believe that soon their attachment to the land that they know so well will be recognised and honoured. When that happens, the Indians of the Chaco will be able to start rebuilding their devastated communities. Meanwhile, the case the Wichí are putting to the world is simple and logical enough:

> Llakha Honhat does not belong to the ahatai, and it would be wrong for them to own it, because it is the home of our ancestors, who lived here for centuries before any ahatai arrived.[6]

Indigenous land rights and the end of 'progress'

Historically, Argentines have prided themselves on their cultural and ethnic affinities with Europe rather than the Latin American interior. Buenos Aires is a self-consciously European city, owing much to Paris and Madrid with department stores modelled on those of London. One aspect of Argentina's obsession with Europe that is rarely discussed is the concomitant refusal to acknowledge the existence of non-European peoples. The African roots of the tango are rarely mentioned in Buenos Aires and the myth that there is no indigenous population obstinately persists. Few *porteños* (as the capital's inhabitants are known) are even aware that the Wichí exist, let alone understand the injustice that has been visited upon them. In this context, the Indians' success in drawing both national and international attention to their land claim is a momentous achievement in itself. The Wichí, although regarded as outsiders by the rest of Argentina, are involuntarily caught up in the wider political struggles of that country. We have already seen that the most vicious attacks on them took place during Argentina's descent into political chaos. In the same way, their entry to the political arena of the 1990s is evidence of a political openness that ten years ago few would have conceived possible. At another level, the question of indigenous land rights can be viewed as a litmus test for the durability of Argentina's commitment to democratic pluralism. So far, they have not been able to force the political system to translate statements of goodwill into practical reality. They have secured from the provincial authorities only a declaration of intent that was almost immediately abrogated.

Wichí political visibility in the Argentina of 'Menemstroika' has not been accompanied by rigid cultural separatism and a retreat into an idealised past. On the contrary, Llakha Honhat has proved adept in its use of Western-style campaigning techniques, including Argentina's recent proliferation of newspapers, television and radio stations. They have also become aware that there is a world outside the Chaco, outside South America, where the question of indigenous land rights looms large. They have learned that there are many tribal peoples throughout the world whose situation is similar to theirs. Like the Australian Aborigines, the Wichí have drawn from their new-found political articulateness the confidence to hold up their way of life as a critical mirror to Western consumer culture.

Indigenous peoples are often described as 'the best ecologists' or 'repositories of ancient wisdom'. These they might well be, but the problem with this form of paternalism in reverse is that it is usually accompanied by a view of indigenous populations as relics of a more wholesome past to be regarded as living museum pieces rather than full citizens. The most admirable feature of the Wichí campaign is that it has accepted help only on its own terms and confronted hostile political institutions on an equal footing without compromising its integrity. Argentina's Indians do not need to use vengeful rhetoric or purple prose to support their quite modest demands. The parlous condition of their territories since criollo colonization expresses better than words the disaster that land enclosure and market economics have been for the Chaco. Wichí campaigners articulate their land claim in terms as applicable to Western rationalist traditions as to their own egalitarian society: the right, as free citizens, to live in and cultivate the land of their ancestors' bones.

[1] This chapter considers the story of the Wichí but there are other indigenous peoples in northern Argentina, such as the Yojwaja, the Toba and the Nivaklé, who have experienced similar persecution by the *criollos*.

[2] Survival International, *Indians of the Americas*, London, 1992.

[3] Rudolf Bahro, *Avoiding Social and Ecological Disaster: the Politics of World Transformation*, Gateway Books, Bath, 1994.

[4] Survival International *Newsletter* No. 33, 1994, p.8.

[5] Ibid.

[6] Ibid.

How sustainable were pre-Columbian civilizations?

Elizabeth Dore

The conquest of America unleashed destruction of human life on a scale unrivalled in history. It may also have been the single most important event to safeguard the plant and animal species of the Americas of the past millennium. In spite of the prevalence of scholarship and myth about a pristine pre-Conquest America, there is increasing evidence that pre-Columbian people were systematically incapable of sustaining the ecosystems upon which their societies depended. Intense environmental degradation may have been a leading cause of the decline of important pre-Columbian civilizations.

The causes of the collapse of the classical lowland Maya have baffled scholars for the past century. Theories abound to explain the demise of the high Mayan cultures around 1000 A.D.[1] Recently, archaeologists have uncovered evidence of acute over-cultivation and soil depletion around the major Mayan sites of Tikkal, Copán and Palenque in the period of the great collapse. Indications of a sharp decline in agricultural yields over a relatively short time point to environmental unsustainability as a prime suspect in the enigma of the Mayan crisis. This view rests neither on biological determinism nor Malthusianism.[2] Rather, it suggests that agricultural methods associated with changing class relations proved inconsistent with the perpetuation of the classical Mayan social order. Although Mayan belief systems around the time of the conquest enshrined the sanctity of nature, it appears that ideology diverged radically from practice. Religious taboo could not prevent ecological change on a scale so vast it undermined the very existence of the classical Maya.

Ecological unsustainability has been suspected as a factor in the crisis of the Aztec Empire as well.[3] New evidence bolsters that interpretation. Examination of soil samples taken from Lake Pátzcuaro, in Mexico, indicates that by the early sixteenth century the landscape of the highlands of Michoacán was seriously degraded.[4] Severe soil erosion, possibly provoking a crisis of food security, may have undermined the power of the indigenous ruling classes, making them vulnerable to conquest. An explicit conclusion drawn from this research is that introduction by the Spaniards of the plough and of cattle grazing, in the past deemed ecologically unsound, may not have been any more environmentally destructive than were traditional indigenous agricultural practices.[5]

Of course, the wave of ecological explanations of pre-hispanic social instability reflects the current popularity of environmentalism. It may be linked, too, to the emergence of ecology as a legitimate academic field, replete with research funding, professional advancement, etc. Nevertheless, to ignore new research, to blindly perpetuate the pristine myth of pre-Columbian America, is naive and romantic. The seductive notion of indigenous peoples naturally living in symbiosis with their environment, then as well as now, is increasingly untenable.[6] Rather than contributing to our understanding of the causes of ecological change, the pristine myth threatens to become a barrier to unravelling the complex dynamics of sustainable and unsustainable development.

Conquest of people, liberation of nature

By the early sixteenth century, the Aztec and Inca Empires were weakened by internal warfare. However, the catalysts of their demise was an exogenous force; the Spanish invasion. Over time European conquest and colonization fundamentally transformed the dominant social and ecological systems of the Americas. From the moment Columbus set foot on Hispaniola, and throughout the colonial era, the appropriation of precious metals was the object of crown rule in America. This brought about a profound ideological revolution in the New World. Even though class relations and technological limitations had severely damaged their environment, nature was sacred to pre-Columbians. For the Spaniards, plunder and extraction of silver and gold was a glorious pursuit.

In the first stage of conquest the imperial enterprise was a ruthless campaign to extract precious metals as quickly as possible from the New World. That quest exterminated the indigenous population of the Caribbean within several decades of contact. Extinction of the Tainos, Caribs and Arawaks – the humans Europeans first encountered in the Caribbean – was caused by disease, hard labour, and social dislocation.[7] Isolated from contact with Europe, Asia and Africa, the native population of the Americas had little resistance to the bacteria and viruses that accompanied Europeans to the New World. Diseases spread rapidly and proved fatal because the Spaniards enslaved the population of the Islands, forcing them to pan rivers for alluvial gold.[8] The colonials terrorised indigenous people into submission by flagellating, maiming and torturing those who resisted their authority. Indifferent to the preservation of the indigenous population, Spanish settlers worked their Indian slaves to death. When the native population of the Caribbean was extinct the colonists continued their plunder on the continent.

By the mid-sixteenth century it was evident that the conquered territories on the American mainland were rich in silver. After royal advisors warned that mines without miners would produce no wealth, preservation of the native population became a priority for the Spanish colonial state to enact measures to prevent a second demographic collapse. But germs and settlers proved more powerful than colonial policy.

Conflict over the immediate interests of the Spanish settlers and the long-term viability of the colonial enterprise first came to a head in the mid sixteenth century. The crown and the conquerors shared an overriding objective: to extract gold and silver from the New World. Individuals sought to accomplish this goal quickly, then return to Spain. The crown, as overseer of the general class interests of the Spanish oligarchy, was concerned with the future prosperity of its colonies. To preserve the labouring class it forbade enslavement of the indigenous population and tried to abolish the *encomienda*, whereby privileged settlers were permitted to collect Indian tribute.[9] Despite the state's intentions, royal decrees were notoriously difficult to enforce in the New World. Consequently, many of the first Indian labourers sent to the silver mines in Spanish America were enslaved in all but name.

As the locus of New World mining changed from looting and panning gold to underground silver mining, there was a fundamental spacial and social reorganization of the continent. Veins of silver were discovered at Zacatecas in New Spain (Mexico), and at Potosí in Upper Peru (Boliva). Supplying the mines with labourers, food, draft animals and timber to build shafts and tunnels altered land use and social relations. The finds were enormous, as were the logistical problems associated with extracting, processing and transporting ore. Cerro Rico at Potosí was particularly rich, yet frustratingly inaccessible, located in the heart of the southern Andes at an altitude of 15,000 feet.

For thirty years, following discovery of the great silver deposits on the continent, individual enterprise in mining continued virtually unrestrained. The age of conquest, from 1492 to 1570, was a period of almost unregulated violence directed against the native inhabitants of the Americas. History repeated itself as *encomenderos* (settlers) imposed draconian working conditions and hours of labour to extract as much ore as humanly possible. As in the Caribbean, this accentuated the mortal effects of disease, exacerbating a population implosion that was on a scale never before or since recorded in the world. Although estimates vary wildly, most scholars think that the population of Ibero-America may have declined from close to a hundred million on the eve of the Conquest to under ten million at its nadir in the seventeenth century.[10]

That demographic collapse safeguarded the ecosystems of the New World for centuries. Early Spanish mining and agriculture altered the environment, certainly. Diverting rivers to pan for gold, felling trees to construct mining shafts, clearing land for cattle grazing and agriculture all caused ecological change. However, their impact on the environment was minor in comparison with the intensive cultivation, foraging and hunting which supported the pre-Conquest population. Not until the nineteenth century did the population of Latin America and the Caribbean approach prehispanic levels. An enduring legacy of conquest for the past five hundred years the region has been under-populated in comparison with Europe, Asia and Africa.[11] As a result there was, until recently, considerably less human pressure on the natural resource base in Latin America than in other parts of the world. This explains, in part, why great forests of the region survived – to be saved at the turn of the 21st century.

*Extract from 'A Socio-Ecological Interpretation of Historical Trends in Latin American Mining', first published in Spanish in *Ecología Política* No. 7 pp 49-68, May 1994.

[1] Theories range from military defeat, internal insurrection, spiritual convulsion, natural disaster such as earth quakes etc. see T. Patrick Culbert (ed), *The Classic Maya Collapse* University of New Mexico Press, Albuquerque, NM, 1973.

[2] Malthus argued that population tends to increase at a faster rate than its means of subsistence and that unless it is checked by a moral restraint or by disease, famine, war or other disaster, widespread poverty and degradation inevitably result.

[3] Michael Harner, 'The Ecological Basis for Aztec Sacrifice', *American Ethnologist* 4:1, pp 117-133, 1977.

[4] Lake Pátzcuarco was in the heart of the sixteenth century Tarascan Empire. Sarah L. O'Hara, F. Alayne Street-Perrott and Timothy P. Burt, 'Accelerated Soil Erosion Around a Mexican Highland Lake Caused by Prehispanic Agriculture', *Nature* (4 March 1993, Vol 362, No.6415, pp48-51; and Karl W. Butzer, 'No Eden in the New World' *Nature* 4 March 1993 Vol 362, No.6415, pp.15-17.

[5] O'Hara et al, op cit.

[6] For a romantic interpretation see Raphael Girard, *Historia de las Civilizaciones Antiguas de América* Ediciones Istmo, Madrid, 1976, 3 vols.

[7] Irving Rouse, *The Tainos:Rise and Decline of the People Who Greeted Colombus*, Yale University Press, 1992.

[8] Alfred W. Crosby, *The Columbian Exchange: Biological and Cultural Consequences of 1492* Greenwood Press, Westport CT, 1987 and *Ecological Imperialism: Biological Expansion of Europe, 900-1900* Cambridge University Press, Cambridge 1986.

[9] The first royal measures to protect the Indians were decreed in 1542 and called The New Laws. Africans brought to the colonies as slaves were less fortunate than the Indians: the New Laws ratified their enslavement. Their financial cost to their owners would serve, in theory, as protection. Henry Stevens, (trans. and ed) *The New Laws of the Indies* The Chiswick Press, London 1893, pp.iii-xvii ff.

[10] Nicolás Sánchez Albornoz, *The Population of Latin America*, translated by W.A.R. Richardson, University of California Press, Berkeley, 1974, pp.37-66; Sherburne F. Cook and Woodrow Borah, *Essays in Population History: Mexico and the Caribbean* University of California Press, Berkeley, 1971, Vol 1, and William Denevan, *The Native Population of the Americas in 1492* 2nd Editions, University of Wisconsin Press, Madison, 1992.

[11] Joan Martínez Alier, presentation at 'Worlds in Collision' Conference, University of Portsmouth, Nov 1992.

Debating 'indigenous' agricultural development: Indian organizations in the Central Andes of Ecuador*

Anthony Bebbington

Advocating the technological quality of indigenous agricultural knowledge and the lead role it ought to play in designing agrarian futures is not a new theme.[1] At the onset of what was to become the 'Green Revolution,' the Rockefeller Foundation hired Carl Sauer as a consultant to advise on how the Foundation should begin a programme of agricultural research in Mexico.[2] Sauer questioned the good sense of rapid technological modernization, arguing that the programme should build on traditional peasant knowledge, drawing on farmers' skills and preserving cultural diversity.[3] But the Foundation saw in Sauer a radical indigenist whose approach would solve neither problems of food supply nor rural and urban poverty.[4] The path it chose was the pursuit of high yield, high input technologies, heavily reliant on chemical fertilizers and pesticides.

While Green Revolution technologies such as those supported by the Rockefeller Foundation have helped increase food production, they have also generated a vigorous debate. The main criticisms levelled at the Green Revolution since the early 1970s are: (1) that agricultural research is concentrated on a few crops and environments at the expense of others; (2) that many modern technologies have been biased against peasant farmers for agro-ecological and socioeconomic reasons; (3) that the introduction of modern technologies has eroded local cultural practice and identity; (4) that the technologies have created ecological problems; and (5) that the technologies have encouraged costly dependencies on imports.[5] These critiques, like Sauer's, have often cast a central role for 'indigenous technical knowledge' in any viable strategy of agricultural developments, on the (claimed) grounds that it is adapted to peasant production conditions, uses no chemical fertilizers or pesticides, is environmentally sound and is culturally appropriate.[6]

But one wonders whether there is a tendency to overstate the potential of traditional knowledge in contemporary political, economic and ecological conditions.[7] Much of the writing on peasant knowledge pays more attention to the creativity of farmers than to the possibility that their actions are influenced by a multitude of acknowledged and unacknowledged conditions and that their actions may have unintended consequences.[8] Such emphases overstate the ease with which a 'farmer first' strategy might be instituted.

This tendency towards cultural romanticism brings us to a further point; for as well as being a means of using the environment, agrarian technology is a symbol that speaks to peasants of their social history and relationships. Indigenous agriculture practised concretely in the field is a sign by which farmers read and understand their identity and their relationship with the past, present and future. Traditional crop practices might then convey both technical significance and a historical and cultural message about continuity with the planting techniques of prior generations. Conversely, as indigenous knowledge incorporates new ideas and material technologies, so it becomes a sign of the group's distance from its past, and its changing relationship to other social groups.[9] New crop varieties produced on agricultural research stations may be a somewhat unfamiliar and alien part of local agrarian technology, a sign that 'the times have changed' and that new bodies of agronomic knowledge are required – 'we can no longer farm in the same way as our grandparents did'. Likewise, when local practices and germplasm are replaced by new technology they become not only a new way of using the environment, but also a sign of new identity: 'look, he plants like a white farmer.'[10]

To plant 'like a white farmer', however, does not necessarily mean one is assimilated into, and subordinated to, a dominant culture – as radical analyses might have us believe. It may well be that for the rural poor, seeing evidence of these modernized practices in their own everyday production strategies could have political implications that are potentially radical – for this can also be a sign to farmers that they are claiming citizenship rights of equal access to technical resources provided by the state, but from which they were previously excluded.[11]

Ecuadorean Indians' demands

is not static. In June 1990, the national umbrella organization for regional indigenous groups, the Confederation of Indigenous Nationalites of Ecuador (CONAIE), called on Indians of Ecuador to participate in 'The First Indigenous Uprising' as a protest against the government's failure to attend to Indian needs.[12] CONAIE's rallying call for the uprising was a list of sixteen points calling the government to protect Indian land rights, to facilitate efforts to strengthen Indian cultural identity, to reorganize market relations to the benefit of indigenous agricultural and artesenal producers, and to push forward an explicit, constitutional recognition of Ecuador as a plurinational entity in which the ethnic rights of indigenous nationalities would be respected.[13]

There are potential tensions between these and other demands of Indian organizations over recent years:[14] tensions between the different Indian concerns for full citizenship rights as *Ecuadoreans*, the right to cultural self-determination, and the demand for a full and fair involvement in Ecuadorean development policies. At times, CONAIE stresses the need to recover and use indigenous crops, technologies, crop-environment theories and cosmologies and to give them a new value as elements of cultural self-determination strategies.[15] Yet, frequently in other parts of the same statement, when demanding equality of citizenship within the Ecuadorean political economy, they call for abstract political rights (of self-determination, of equal political participation etc) but also for equal access to specified aspects of economic and rural modernization: agricultural technology, credit, marketing facilities, product processing etc. These latter demands imply that Indians should be able to produce and sell in the same way that non-Indian farmers have done for many years.

These tensions in the national Indian movement point to the difficulty of defining and sustaining an indigenous identity in an economically modernizing, fragile democracy; of combining being 'indigenous' with a search for 'development and democratization', of combining tradition and change, and of bringing the past into the present. 'Tensions' are not, however, insoluble contradictions. There is another demand stressed by the Indian movement that holds the key to the solution of these contradictions: CONAIE claims that indigenous people, being at once Indian and Ecuadorean, have the right to self-government and self-management within their communities, as well as the right of access to state resources for use within those communities.[16] As I shall argue, it is in such initiatives for a 'bottom up' self-management of the modernization process, far more than in the recovery and perpetuation of technological and cosmological traditions, that many local Indian groups are already building the basis of their own agrarian strategies, regardless of CONAIE's or government postures. Similarly, it is in these initiatives that they simultaneously address technological, cultural and organizational concerns, regardless of the development literature's tendency to separate them. Let me illustrate this approach with reference to one part of the central Andes of Ecuador.

Indian struggle and agrarian change

The province of Chimborazo was one of the last bastions of the traditional large rural estates, or *haciendas*, in Ecuador.[17] By the mid-twentieth century the haciendas in the north of the country were modernizing, selling marginal lands to indigenous people and substituting capital for labour, but the haciendas of Chimborazo invested little in their agriculture.[18]

Under the hacienda system, proportionally little land was dedicated to cash and food crops, and the remainder was pasture.[19] While this allowed a rotation with long fallow periods, by the mid-twentieth century some agrochemicals and new crops were introduced into crop-production techniques. For many indigenous farmers in the region, their first experience of modern technology was to see it when working on these haciendas, or when working as migrant labourers on modernized large farms elsewhere. In either case, they experienced it as a technology that increased yields and that was available to the white *hacendados* (hacienda owners), but not to the indigenous farmer. It was in this sense a technology associated with social power. The Indian production system linked to the hacienda was based upon small plots of land dedicated to a wide range of traditional crops grown primarily for family food. Agrochemicals were rarely used. Their's were agro-ecologically sound practices, dependent on heavy inputs of manure from animals kept on the hacienda's pasture lands.[20]

By the 1950s, the seeds of decline for this set of social relationships were being sewn both locally and centrally. The local pressures came from an increasingly organized indigenous peasantry initiating intensified levels of agrarian conflict as they sought to recover the land that four hundred years previously had belonged to their ancestors. At the same time, the central state saw the haciendas' resistance to change as an obstacle to national modernization. The levels of Indian and peasant mobilization in Chimborazo were such as to cause concerns about internal security. Agrarian reform was implemented in 1964 and 1973, and in response to the levels of conflict in Chimborazo, the province was declared a priority zone for the application of the 1973 legislation.[21]

From being a region dominated by haciendas, Chimborazo saw the country's most thoroughgoing application of the agrarian reform legislation. This brought about profound changes in agricultural conditions. The sub-division of the hacienda among indigenous farmers meant that the hacienda's pasture lands that were previously used by indigenous farmers were turned to crops. Farmers also sold animals to purchase their plots. The combined effect was that old organic fertilization strategies became increasingly problematic: fewer pastures, fewer animals, less manure. Also, in the absence of other local income opportunities, an increase in local populations has led to the progressive subdivision of land and reduction of fallow periods. This has resulted in an impoverishment of soils and a decline in yields. Poorer soils have meant greater plant susceptibility to diseases, and farmers have turned to synthetic fungicides to combat these.

Accompanying this has been greater contact with the market through the sale of products. This discouraged the cultivation of many traditional Andean crops for which there is little market, and fostered a move to non-traditional crops which do have a market. A further development has been the intensification of seasonal migration to urban and coastal areas.

Although it is a necessary survival strategy, indigenous people associate this periodic migration with social problems and weakened cultural practices in their communities. Some of these problems, such as petty theft, bad manners and petty violence are perceived as urban pathologies imported into communities, undermining indigenous forms of life. Others are the undesirable by-products of migration: weaker family ties, marital conflicts, health problems and declining levels of participation in and support for communal activities. All these material and social changes have set limits on what is perceived as a possible and desirable future for indigenous agriculture.

The social and administrative changes initiated by the agrarian reform are also significant. Not only did agrarian reform signify an attempt by the state to assume a key role in the agricultural development of the region, but it also eased the entry of religious and non-governmental organizations, and with this a diversity of perspectives on the nature of Indian identity, and of appropriate indigenous agrarian practices.

Indigenous conceptions of the future

Today, one of the main points of technological debate among and within Ecuador's indigenous federations is the extent to which they should work with and disseminate modern agrochemical and crop technologies among their members. The three 'developmentalist' organizations (AIECH-San Lucas, AOCACH and UOCACI) operating in an area where the subdividing of smallholdings, land degradation and periodic migration are particularly advanced and where evangelical Christianity is especially influential, favour a role for modern technology in indigenous agriculture, believing it can make an important contribution to improving rural income and nutritional levels. This is perceived as a necessary technological response to the serious grazing crisis and land degradation that have left much land in dire need of nutrients from outside the ecosystem (such as those found in chemical fertilizers). In spite of this affirmation of modern technology, the developmentist organizations place high value on the knowledge of indigenous farmers. This is not, however, a validation of prior 'traditional' agrarian practices, but rather of the technical creativity of indigenous farmers, and their capacity for management and administration.

At the other end of the spectrum are the more radical federations such as UCASAJ which are more sceptical of the role of modern technology in Indian agriculture, and more overtly critical of the capitalist character of the market system which they perceive as being the same set of relationships as has brought modern technology to the region.[22] UCASAJ have sought to recover and stimulate the production of Andean crops, the use of organic fertilizers, and the replacement of agrochemicals. However, the existence of greater market opportunities for certain crops, the problem of soil degradation, and the consequent pressures from their members have led them in the last few years to work with agrochemicals, focusing instead on trying to limit input use and teach greater care in the use of the products.

Common to both types of organization is a constant concern to relate the methods and technologies with which they work to their ultimate goal of increasing indigenous participation and control over the process of agrarian change. In this regard there are two recurrent strategies. One is to set up seed and fertilizer distribution systems managed by a local indigenous organization, and regulated through periodic inter-community meetings. A second activity has been to train indigenous agricultural promoters who are then controlled by the communities. These promoters are trained both by the organizations' own agronomist and by sending them on courses at formal institutions. Once trained, they function like any state agricultural advisor, except that their remuneration is lower, and they do not work full time.

For the organizations seeking to recover traditional indigenous technologies, the strategy is justified as a rejection of white and capitalist culture, and as an affirmation and validation of indigenous identity.[23] It is also seen as a way of reducing dependence on the market, as well as the costs of production and environmental pollution. These are the same justifications for recovering traditional technologies that one finds in agroecological texts in the North.[24] However, in many cases, the membership of these organizations has not accepted these justifications on the grounds that retaining traditional technologies in market oriented production strategies has proved very difficult. What is interesting is that in response to members' concerns, UCASAJ has adapted its strategies accordingly.

The justifications given by 'developmentalist' indigenous organizations for strategies based on selective modernization appears much more in harmony with the attitudes and practices encountered among their members. Modernization, far from being a cause of cultural erosion, is seen as a means of cultural survival. With periodic migration seen as the main threat to cultural coherence, technological modernization, along with the promotion of non-traditional cash crops, is justified as a strategy for increasing local income opportunities and so reducing pressures on migration.

Farmers see a close relationship between rural residence and their identity as indigenous farmers. Rural residence, which, in turn, strengthens their social organization, is thus seen as being more important than retaining traditional technologies. Yet, in current conditions, productive strategies based on earlier, non-modernized technologies do not appear a viable means of ensuring rural residence. The implication is that indigenous culture must constantly adapt in order to survive, and to sustain group cohesion and forms of self-management.[25] There are also symbolic reasons why farmers use modern technologies. Indigenous farmers frequently associate 'traditional' technologies with the subjugated forms of living associated with the hacienda and wish to distance themselves from past agrarian practices. Traditional technology thus speaks of social relations to be rejected. Furthermore, embracing modern technologies is to make a statement about social equality and the demand for access to traditionally white technology. The use of modern technology is thus part of the wider quest for citizenship rights.

Behind the apparent differences in the viewpoints of the organizations are a set of more significant shared rationales revolving around the concern for local self-managed strategies in which local groups define their own agricultural goals. First is the attempt to control the way resources for development projects are used in their territory, adapting them to the needs of their membership, and demonstrating Indian ability to use and manage modern administrative methods, and (in the case of the developmentalist organizations) modern technologies. Secondly, there is a concern to strengthen indigenous organizations as a vehicle for demanding social changes to improve Indian production (the strength of support for the uprising among these organizations suggests this goal has been quite successful).

It is interesting that the work methods of the indigenous organizations are very similar to those of the state's and NGOs' agricultural development programmes. This is both a weakness and a strength. The weakness is that the similarity of style leads the bases to criticize the projects on the same criteria as they have criticized state initiatives, even though the organizations tend to reach farmers unattended by state services. The strength is that, precisely because they are assessed in the same way, other qualities of the projects take on that much more significance. The fact that most of the administration is done by indigenous people selected by their communities, and that while the organizations are not perfectly democratic, their projects are more responsive to the preoccupation of local farmers than are the state's because the leaders are more easily held to account, all mean that the traditional ethnic blindness of state projects becomes that much more obvious. The implication is that indigenous people should have a prominent role in all rural development planning and implementation.

This is being strengthened by an additional political development: the increasing indigenous control of rural space. Indigenous people are now being elected to local government positions in the local levels of the Ecuadorean state,[26] and as white and mestizo presence in rural areas declines every year, it seems that rural areas are being returned to indigenous people as a space in which to practice indigenous culture and agriculture, and to take over rural administration more generally. The Indian movement, locally and nationally, has stressed that these tendencies do not reflect a desire to slide into self-sufficiency and separatism. Indeed the very idea of claiming the right of access to modern technology and the resources that underlie it, implies maintaining relations with the state and formal agricultural scientists. Rather, the point is that the terms on which this access is to be gained should be set by the indigenous organizations. The indigenous agricultural development presaged in these federations explicitly involves changing existing socio-political relationships and the way the state is organized.

Conclusion

An indigenous strategy will require access to modern technologies, knowledge and resources, and will not be entirely native (although it will be locally controlled).[27] 'If other social groups have not had to be self-reliant, why should indigenous farmers be expected to be so?' goes local talk. Successful appropriation of such modern technologies will require scientific skills that farmers do not have, and an indigenous agricultural strategy will be firmly integrated in a wider political economy, incorporating modern technologies and cash crops.

But such political economic integration will not necessarily lead to dependence and cultural assimilation. Rather it involves the incorporation of non-local resources and ideas into the strategies for surviving as an *indigenous* farmer. Thus, indigenous farmers are now drawing upon the resources they have managed to secure by demanding their rights as Ecuadorean citizens. But a key aim of farmers' strategies is to control and administer the resources they are claiming from the Ecuadorean state, through their organization. An agrarian strategy that is indigenous is also coming to mean a strategy in which local organizations administer resources and processes of development in rural space.

*Extracts from Anthony Bebbington, Working Paper No. 45, Centre of Latin American Studies, University of Cambridge, 1992.

[1] A.J. Bebbington & J. Carney, 'Geographers in the International Agricultural Centers. Theoretical and Practical Considerations' *Annals of the Association of American Geographers* 1990, 80(1): 34-48.
[2] B.H. Jennings *Foundations of International Agricultural Research: Science and Politics in Mexican Agriculture*, Boulder, Westview Press, 1988.
[3] C.O. Sauer. Letter to Rockefeller Foundation, February 10. 1941. Rockefeller Foundation/ reference number:1.2/3.2, p.2, Rockefeller Foundation Archives, Tarrytown, NY, cited in Jennings 1988 op cit; N. Entrikin, 'Carl O. Sauer, philosopher in spite of himself' *Geographical Review*, 74, 1984, 387-408.
[4] Jennings 1988, op cit.
[5] M. Altieri, *Agroecology. The Scientific Basis of Alternative Agriculture*, Westview, Boulder, 1987; S. Biggs & J. Farrington, *Agricultural Research and the Rural Poor: A Review of Social Science Analysis*, International Development Research Centre, Ottawa, 1991; P. Blaikie, 'The Theory of Spatial Diffusion: a Spacious Cul-de-Sac' *Progress in Human Geography* 1987, 2, p.268-295; A. De Janvy, *Land Reform and the Agrarian Question in Latin America*, John Hopkins University Press, Baltimore, 1981; B. de Walt, 'Halfway There: Social Science in Agricultural Development and Social Science of Agricultural Development' *Human Organization* 1988, 47, pp. 343-353; K. Griffin, *The Political Economy of Agrarian Change*, Macmillan, London 1974; C. de Hewitt Alcantará, *Modernizing Mexican Agriculture*, United Nations Research for Social Development, Geneva, 1976; L. Yapa, 'The Green Revolution: A Diffusion Model', *Annals of the Association of American Geographers* 1977, 67, pp.350-359.
[6] For example, Altieri 1987, op cit; R. Chambers, R. Pacey, A. and L.A. Thrupp (eds) *Farmer First, Farmer Innovation and Agricultural Research*, Intermediate Technology, London, 1989; P. Richards, *Indigenous Agricultural Revolution, Ecology and Food Production in West Africa*, Hutchinson, London 1985 and *Coping with Hunger. Hazard and Experiment in a West African Rice Farming System*, Allen & Unwin, London 1986.
[7] P. Blaikie & H. Brookfield (eds) *Land Degradation and Society*, Methuen, London 1987.
[8] A. Giddens, *Central Problems in Social Theory*, MacMillan, London 1979; D. Gregory, 'Suspended Animation: the Stasis of Diffusion Theory' in D. Gregory and J. Urry (eds) *Social Relations and Spatial Structures*, MacMillan, London, 1985.
[9] L. Busch, 'On Understanding Understanding: Two Views of Communication', *Rural Sociology* 1978, 43(3),pp.450-473.
[10] A.J. Bebbington, 'Indigenous Agricultural Knowledge, Human Interests and Critical Analysis: Reflections on the Role of Farmer Organizations in Ecuador', *Agriculture and Human Values*, 1991, vol 18(1-2), pp.14-24.
[11] A.D. Lehmann, *Democracy and Development in Latin America. Economics, Politics and Religion in the Postwar Period*, Polity Press, Cambridge, 1990.
[12] L. Macas, 'El levantamiento indígena visto por sus protagonistas' in ILDIS, *Indios*, Quito, 1991, pp.17-36. Luis Macas was president of CONAIE in 1991.
[13] MLAL-Abya-Yala, *KIP.U: El Mundo Indígena en la Prensa Ecuatoriana*, vol 14, Quito, 1990.
[14] CONAIE *Nuestro Proceso Organizativo*, Confederación de Nacionalidades Indígenas de Ecuador, 1989.
[15] CONAIE 1989, op cit.
[16] Macas, 1991, op cit pp.25-26.
[17] E. Haney & W. Haney, 'The Agrarian Transition in Highland Ecuador: from Precapitalism to Capitalism in Chimborazo' pp.70-91 in W. Thiesenhusen (ed) *Searching for Agrarian*

Reform in Latin America, Unwin Hyman, London 1989.
[18] O. Barsky, *La Reforma Agraria Ecuatoriana*, Corporación Editorial Nacional, Quito, 1984; P. Sylva, *Gamonalismo y Lucha Campesina*, Abya Yala, Quito, 1986.
[19] E. Haney & W. Haney, *La Transformación de la Estructura Agraria en la Provincia de Chimborazo*, Land Tenure Center, University of Wisconsin, 1984.
[20] A. J. Bebbington, *Indigenous Agriculture in the Central Ecuadorian Andes. The Cultural Ecology and Institutional Conditions of its Construction and its Change*. Ph.D. dissertation. Graduate School of Geography, Clark University, Worcester, 1990.
[21] Haney & Haney 1984, op cit; Sylva 1986, op cit.
[22] UCASAJ, *Pueblo Indio*, Unión de Cabildos de San Juan, Chimborazo, 1989.
[23] MICH, 'Movimiento Indígena de Chimborazo, MICH' in CONAIE *Nuestro Proceso Organizativo*, 1989, op cit, pp.195-202.
[24] For example, Altieri, 1987, op cit.
[25] V.G. Ramón, *Indios, Crisis y Proyecto Alternativo*, Centro Andino de Acción Popular, Quito, 1988. Ramón's example is of 'the initiatives of the Otavaleños [commercial textile producers] or the onion producers of Cangahua, who, on the basis of controlling land resources or artesenal production, have developed marketing networks, and increased their proportion of value added in production, which in turn enables the sustained reproduction of the ethnic group,' pp.24-25.
[26] J. Sánchez-Parga, *Faccionalismo, Organización y Proyecto Etnico en los Andes*, Centro Andino de Acción Popular, Quito, 1989.
[27] J. Blauert, 'Autochthonous Development and Environmental Knowledge in Oaxaca, Mexico' in P. Blaikie & T. Unwin (eds) *Environmental Crises in Developing Countries*, Monograph # 5, Developing Areas Research Group, Institute of British Geographers, 1988.

Oil, lawlessness and Indigenous struggles in Ecuador's Oriente

Judith Kimerling

In 1967 the US-based company, Texaco, discovered commercial quantities of oil in Ecuador's Amazon region, 'the Oriente'. The oil boom that followed, and continues today, is the primary engine of change, conflict and environmental degradation in the region. Despite the recent proliferation of international agreements to protect the environment and extensive environmental rights and obligations under Ecuadorean law, the rule of law has not reached the oilfields. In recent years the industry and the Ecuadorean government have appeared more sensitive to the growing national and international criticism of environmental and human rights conditions. However, this has not yet been translated into significant improvements in the field. Government and corporate officials have not opened their actions to public scrutiny nor have they been held accountable for the injuries they have caused.

In theory, Ecuador is a constitutional democracy, having reestablished democracy in 1979 after seven years of military rule. In practice, however, democratic institutions remain fragile and underdeveloped, and a strong executive dominates the government. The courts have become increasingly politicized, inefficient and corrupt. A popular saying, 'the law is only for those with the ponchos,' reflects the general belief that only the most marginal citizens – the indigenous peoples – are not above the law. According to most estimates, indigenous peoples make up some forty per cent of Ecuador's population. Ecuadorean society, however, is characterized by deep racism against its indigenous peoples, widespread poverty, and extreme inequality. This social reality supersedes the legal guarantees and democratic ideals written in the constitution and other laws. Political and economic power is concentrated in the hands of a small elite, and the majority of Ecuadoreans are excluded – in practice if not by law – from meaningful participation in the political system. Most governmental decision-making is characterized by secrecy and accountability of public officials is limited.

The indigenous peoples of the Amazon live far from the centres of power. Poor roads and transport services and the near-absence of radio and telephone communications augment the geographical distance. Cultural, linguistic and historical distances further separate Amazonian peoples from their government. The government views the Amazon as a frontier to be conquered – a source of wealth for the debt-burdened state and an escape valve for demographic and land distribution pressures.

Government policies generally seek to develop and colonize the Amazon, and promote the assimilation of indigenous peoples into the dominant Ecuadorean culture.

The oil boom reflected and reinforced two tiers of inequality. Ecuador's dependence on the transnational company Texaco to transfer petroleum technology and finance and operate the infrastructure that turned the country into a petroleum exporter reflected a 'First World-Third World' inequality. Within Ecuador, the remote indigenous homelands that host the oil reserves are effectively a 'Fourth World'.

Amazon crude and state policy

Initially, the boom brought unprecedented economic expansion but in the 1980s per capita income declined dramatically. Ecuador has amassed an enormous debt burden and remains dependent on foreign capital and technology to locate and develop most of its oil reserves. In addition, the benefits of the boom have not been well distributed. The government estimates that 67 per cent of the population lives in poverty, compared to 47 per cent in 1975. In the Oriente, oil development has created poverty among indigenous peoples by destroying natural resources used for nutritional, medicinal, domestic, artisanal, religious and recreational purposes, by endangering food and water supplies, and by violently disrupting their way of life. Meanwhile petroleum accounts for roughly forty to fifty per cent of Ecuador's export income and national budget, depending on the international price of oil.

Under Ecuadorean and international law, the state is charged with environmental protection responsibilities. Nonetheless, Ecuador's governments have shown little or no willingness to comply with the law. Instead of assuming an authoritative regulatory role in environmental affairs in the oilfields and protecting the rights of residents, the government has essentially behaved like a business partner to the industry. The government has financial incentives to keep environmental protection costs to a minimum as it reimburses the production and exploration costs of operators developing commercial reserves. In addition, many officials fear that if environmental protection becomes costly, investment will go elsewhere. Oil companies are virtually self-regulating in environmental matters and standards are predominantly determined by the companies' internal policies rather than by the rule of law. Even when legal requirements are precise, such as the prohibition on dumping oil on roads, routine violations are overlooked and the law is not enforced. The failure of the government to establish meaningful environmental controls lies in stark contrast to its dogged regulation of economic aspects of petroleum development, including production rates, and its active participation in exploration and production activities.

In practice, government intervention in the petroleum sector is dominated by two powerful agencies: the Ministry of Energy and Mines (the Ministry) and Petroecuador, the national oil company. Ecuador does not have an independent ministry to protect the environment or indigenous rights. The small environmental units that were created in the Ministry in 1984 and in Petroecuador in 1990 have been ineffectual, operating more as advisory bodies than regulatory agencies. A semi-autonomous company with several subsidiaries, Petroecuador behaves like both an oil company and a government agency. It was first set up as CEPE in 1971 in the wake of Texaxo's oil discovery and was reorganized as Petroecuador in 1989 to assume operation of the Texaco facilities and expand state exploration and production activities. (Texaco's production contract with the government expired in 1992.) It is fair to say that Petroecuador is the stepchild of Texaco and the Ecuadorean state. Moreover functionaries in both Petroecuador and the Ministry have basically been trained in the Texaco school. The current chief of the environmental unit in the Ministry of Energy and Mines, for example, used to work for CEPE as a contract employee in Texaco. Now he is an employee of Petroecuador on loan to the Ministry. Texaco's training did not include environmental planning, monitoring or controls.

Environmental conditions at the sites that Petroecuador inherited from Texaco – and at the sites developed by Petroecuador using the technology it acquired from Texaco – are so appalling that some of the foreign companies comprising the post-Texaco generation of operators go to great lengths to distinguish their practice from Texaco's. Since 1990 the government has formally recognised the need for environmental improvements in the oilfields. However most functionaries charged with environmental protection duties do not live in the Oriente and are primarily petroleum engineers, who became environmentalists overnight. Many officials in the Ministry and Petroecuador still do not believe in environmental protection. One told me 'petroleum isn't really harmful; people just think it's dangerous... because it's dark-coloured; like [they think of] the Palestinians.'

It is not easy to awaken environmental awareness about the dangers of irresponsible oilfield practices in a nation where the first barrel of commercial oil was paraded through the capital like a hero, and where oilfield workers in the 1970s and 1980s believed that Amazon crude had medicinal properties, and brought jars of it home from the Oriente, to give to parents suffering from arthritis. Similarly, many workers believed that it could slow or prevent balding. They periodically covered their hair and scalp with crude, and either sat in the sun or covered their heads with plastic caps overnight. To remove the crude, they washed their hair with diesel.

Environmental impact

Adverse environmental impacts occur at every stage of oil development. The most acute impacts are caused primarily by contamination and deforestation. Some deforestation results directly from oilfield activities; and massive deforestation and wildlife destruction typically occur when road construction to service commercial fields opens previously inaccessible forest to land speculation, colonization, ranching, logging, hunting and agro-industry. Contamination is caused by erosion from cleared areas, accidental spills of oil and other chemicals, routine discharges and emissions of wastes with toxic constituents, sloppy maintenance procedures, and noisy motors and other machinery.

Texaco drilled 339 wells in an area that now encompasses 442,965 hectares. Some 235 are active, generating over five million of gallons of toxic wastes every day. These wastes contain hydrocarbons, heavy metals and toxic levels of salts, and most are discharged into the environment without treatment or monitoring. In addition, toxic drilling wastes, and wastes from maintenance activities at wells, flow lines and other facilities, have been dumped into the environment without treatment or monitoring. Most of the wastes have been dumped into open holes in the ground, known as 'pits'. Typically, the pits are unlined, allowing toxic leachate to filter into the groundwater. Rainwater freely enters the pits, and becomes contaminated by the wastes. When open pits fill up, they overflow, contaminating surface waters and soils. To limit overflows, waste pits at production sites, where wastes are generated on a ongoing basis, are drained by an outfall pipe. Depending on the level of wastes in the pits, the outfall discharges untreated liquid wastes or heavy crude into the environment.

In addition, oily wastes are regularly applied to most oilfield roads for dust control and maintenance purposes (though Occidental now uses water), and wash into the environment with the frequent rains. Most of the gas that is extracted with the oil is burned as a waste, without environmental controls or monitoring, wasting a precious resource, and contaminating the air with greenhouse gases and precursors of acid rain (nitrogen oxides and sulfur dioxide), among other contaminants. As well as routine discharges, spills from pipelines, wells, tanks and other facilities have occurred with regularity. Between October 1994 and October 1995, Petroecuador recorded 58 spills; over half were attributed to corrosion in ageing flow lines. During the time that Texaco operated the main trans-Ecuadorean pipeline, accidental spills from that line alone sent an estimated 16.8 million gallons of crude into the environment, mostly in the headwaters of the Amazon basin. Texaco did not develop a spill contingency plan and spills were not cleaned up.

Countless rivers and streams that supply water and fish to Oriente communities have become severely contaminated by spills and discharges. Many waters that were once rich in fish now support almost no aquatic life. Other animals, crops and forest flora have also been killed by pollution. Increasingly, both indigenous and settler residents attribute a number of health problems to the pollution that saturates the area – including malnutrition, skin rashes, headaches, fevers and miscarriages.

Beginning in 1992, waste pits at dozens of well sites were covered with dirt, without testing, treating, or removing the wastes. A second type of 'cleanup' was carried out at numerous sites that are contaminated by oil spills or overflow from waste pits. Crude oil and contaminated soils and vegetation were buried in small unlined holes in the ground, again without testing, treatment or monitoring. In the holes, the oil is less visible – for now – but it will continue to contaminate the environment. At some sites, covered wastes have already oozed up, through the dirt, to the surface.

Legal action

In 1993 a $1.5 billion class action lawsuit was filed in federal court in New York against Texaco, on behalf of indigenous and colonist residents who have been adversely affected by the company's Oriente operations. They are accusing Texaco of negligence, public and private nuisance, strict liability and trespass. The plaintiffs sued in New York because Texaco's corporate headquarters is there, and they allege that a substantial part of the company's harmful acts and omissions – directing the Oriente operations – took place there.

Having received extensive coverage in the Ecuadorean press, the lawsuit has generated enormous interest among Amazonian organizations, who are hopeful that authorities in the US will help them. However the suit is not well understood by most residents, and the attorneys who filed the suit have resisted initiatives that seek to coordinate the litigation with local organizations, preferring instead to work on their own or with selected individuals (full disclosure: I have worked on these initiatives with the Federation of Comunas Union of Natives of the Ecuadorean Amazon, FCUNAE, the Indigenous Organization of the Cofán Nationality of Ecuador, OINCE, and Organization of the Huaorani Nationality of the

Ecuadorean Amazon, ONHAE, and some indigenous plaintiffs). Meanwhile the Ecuadorean government has vigorously opposed the lawsuit, arguing that litigation in US courts would be a 'serious disincentive' to US companies to invest in Ecuador, and 'could do severe harm to the Republic of Ecuador'. Initially, it presented a diplomatic protest to the US Government, requesting it to advise the Court that the case should be heard in Ecuadorean rather than US courts. When reported by the press in Ecuador, the protest provoked a storm of criticism, and charges that the government had sided with a foreign corporation against its own citizens.

The US government did not act on the protest, prompting the Ecuadorean government to hire a New York law firm to present a 'friend of the court' brief in support of Texaco's efforts to dismiss the lawsuit. The Court stated that the 'exercise of judicial jurisdiction over events initiated in the United States and carried out abroad (whether in Ecuador or elsewhere)' would be 'country-neutral' and 'cannot encourage or discourage investment in any particular country'. The Court is expected to decide whether to accept the case in 1996.

Clean-up or cover-up?

In what is widely regarded as an effort to undermine the litigation and repair their tarnished international images, Texaco, Petroecuador and the Government (through the Ministry) have negotiated a series of agreements, which they claim will address environmental issues raised by the suit. Publicly, and repeatedly, they have announced that Texaco will clean up damaged areas. However, they have refused to release the agreements – and the environmental audit that preceded them – to the public, or provide residents and environmentalists with details about how and when the clean-up will be carried out. The secretive nature of the negotiations, and a concern that the purpose is to cover up rather than clean up, prompted the Catholic Mission in Coca to refuse to accept a plane that would have been given by Texaco to the Mission, for use as an ambulance, as part of the agreement.

When the Government finally released some agreements to the Environment Commission of the National Congress (which passed them on to environmental groups) in 1995, environmentalists' worst fears were confirmed: no action will be taken at the 'clean-up' sites mentioned above; and no action will be taken to assess and remedy air pollution, or contamination of ground and surface waters. At sites that are targeted for clean-up, chemical sampling and monitoring are inadequate; the levels of contamination that will be tolerated in water and soils are unreasonably high; and clean-up technologies and procedures are vague and imprecise.

Because oilfield practice in the Oriente lags far behind accepted practice in the industrial world, some foreign oil companies have responded to international criticism and grassroots pressures by directing their Ecuadorean subsidiaries to implement environmental measures that go beyond what is required of them legally. But there is no independent oversight or monitoring. Even the most innovative companies have not matched their sweeping claims – that they can extract oil from the Amazon without seriously harming the environment – with the information or the transparency that is needed to verify their claims. Instead, environmental decision-making and monitoring are shrouded in secrecy without meaningful participation by affected residents. Vital information is withheld from the public (even though the legality of this practice is dubious). As a result, the effectiveness of corporate environmental initiatives has not been demonstrated. In some cases, it is unclear whether companies' modified practices yield a net environmental benefit.

For example, some companies are now using directional drilling technology to minimize clearing of the forest. This allows up to twelve wells to be drilled from a single platform, instead of clearing a separate platform and access road for each well. There is no question that this will be an aesthetic improvement if colonization around the wells can be controlled. However, directional drilling typically generates larger quantities of drilling wastes than vertical drilling. Serious concerns have been raised about the disposal of those wastes by all Oriente operators, throwing into question the net environmental benefit of directional drilling, as implemented.

Maxus, the operator of new production facilities in indigenous Huaorani territory and Yasuni National Park, has collaborated with Jatun Sacha, a local foundation with ties to the Missouri Botanical Garden, in a highly-publicized 'reforestation' project along its new road. Although Maxus claims the project uses native species, a report halfway through the project revealed that only ten hectares had been reforested with some forty species of native trees, while eighty hectares were revegetated predominantly with a fast-growing grass, native to Africa (Brachiaria decumbens). Reportedly, the reforested areas are not growing well; so conceivably, the benefit gained in diminishing erosion and sediment-loading outweigh the risks of introducing this foreign species into the area. The key point is that there has been no independent review of the project.

The need for public review is also evident in some companies' plans to re-inject certain liquid production wastes underground. While reinjection to an environmentally-safe depth is generally regarded as the best environmental practice for those wastes, injection zones must be properly selected and activities carefully monitored, in order to prevent injected wastes from contaminating underground aquifers or finding their way to the surface. Oilfield workers report that the ARCO company is having difficulty locating suitable injection zones because its area of operations crosses a major fault, and is so geologically dynamic that the ultimate destiny of underground wastes cannot be assured. The life expectancy of the injection zones currently used by Maxus is reportedly no more than three years, after which the formations are expected to become saturated and new ones will be needed.

These private initiatives are a dramatic departure from the corporate indifference and environmental recklessness that characterized past operations – and continues to characterize most current ones. However, they are neither uniform nor complete. Most importantly, they are not open to public scrutiny and environmental protection should not depend exclusively on the goodwill of corporate officials.

Indigenous peoples and environmentalists take action

Legalization and demarcation of indigenous territories has long been the frontline of defence used by indigenous organizations to protect indigenous peoples from the oil boom. Some clans and communities have also taken direct action against the companies – occupying work sites, blocking roads, or raiding oil camps. Usually, these actions end without bloodshed and the underlying tensions are defused by a combination of threats and promises from the government to negotiate further, or by company support for local infrastructure or other 'gifts'. In a few cases, residents have forced companies to re-route their operations, protecting small islands of territory in the heart of their communities. For example, the Cofán people prevented Texaco from extending a road and drilling additional wells in Doreno, and the Quichua stopped ARCO from completing some exploratory seismic studies in Sarayacu. Both companies, however, continued to expand their activities into nearby areas. In addition, the Tagaeri Huaorani – who were dislocated from their ancestral lands by Texaco and have violently resisted efforts by missionaries, oil companies, adventurers and other Huaorani to contact them peacefully – temporarily stalled exploratory activities by Petrobras in the 50,000-hectare forest that is currently their home. That work, however, is expected to resume in 1996, despite fears that any encounters with the Tagaeri will be violent, and could result in the extinction of the group. (The Petrobras consortium includes at least one European company, Elf Aquitaine).

In response to the surge in international concern for tropical forests, Oriente communities and organizations have also begun to appeal to world opinion, by documenting and denouncing harmful operations. Fieldwork

I conducted with FCUNAE in 1989-90 and the publications based on that work helped to demonstrate the power of information and effective documentation, both written and photographic. The work is widely credited with attracting national and international attention to previously-unheeded complaints by residents about environmental conditions in the oilfields. Local organizations have also begun to appeal to public opinion for support when they advance alternative development initiatives, and for protection when they risk bloodshed and repression by taking direct action.

Although the Tagaeri is the only group that is still at war with the oil companies at this time, most Oriente communities do not want the industry to work in their lands. The government, however, has repeatedly rejected proposals by indigenous organizations for temporary moratoria on new development activities and for better controls based on investigations of oilfield operations by independent groups that include international experts and representatives of residents. Decisions as to where, when and how to carry out oil development activities rest exclusively with the government and the companies. Recent land titles ceded by the government to indigenous peoples provide that title can be terminated if residents 'impede or obstruct' oil or mining activities. Most Oriente peoples believe they do not have a choice as to whether oil companies may enter their territory, and that the government and the companies will use force if they try to stop them.

As a result, most communities and organizations now seek to negotiate directly with the companies, in order to mitigate adverse environmental and social impacts and share in the benefits of development. When pressed, most companies agree to fund some community infrastructure or give other 'gifts' to residents; however, they strongly resist negotiating environmental protection measures. Because the negotiations are carried out under threat of entry without compensation, residents generally believe the only choice they have is to take what they can get, or get nothing, as in the past. Most residents have growing needs for cash income, and compelling education, health care and community development needs.

Increasingly, local and environmental critics charge that companies buy off selected individuals or communities cheaply, creating or exploiting divisions among residents and making promises they do not keep; that they attempt to substitute relatively minor community development works for serious environmental protection; and that they use selected residents as a shield against criticism by environmentalists who are reluctant to appear to question the right of residents to negotiate their own destiny.

One group that has made some progress in monitoring environmental impacts and participating in negotiations is the Organization of Indigenous Peoples of Pastaza (OPIP). OPIP represents Záparo and most of the Quichua who are affected by ARCO's activities, and led a historic march to Quito in 1992, that resulted in legalization by the government of 1,115,574 hectares of indigenous territories. The group garnered local and international support to compel ARCO to negotiate environmental, community development, and long term resource management issues. However, after a promising start, the dialogue was suspended, over OPIP's objections.

The Cofán have combined efforts to negotiate with unrelenting nonviolent action. The Cofán number roughly 700, and are on the edge of extinction as a people. The first commercial strike of Amazon crude was in Cofán territory; as a result, the Cofán were dispossessed of huge tracts of their traditional territory. They now live in five small, non-contiguous pockets of land. One community is in the Cuyabeno Wildlife Reserve, where the Cofán have developed a thriving ecotourism business. In OINCE the Cofán have joined together to protect what remains of their territory and way of life. With support from a powerful neighbouring ecotourism operator and environmentalists, the Cofán have led efforts to protect what remains of Cuyabeno from illegal oil exploration by Petroecuador. As unpaid park guards, they have monitored Petroecuador's activities and insisted that the company comply with the law. At times, they have taken direct action to halt illegal activities, while convincing many of the workers and military officials they encounter – as well as press and the public – of the justice of their claims, and the injustice of the actions they challenge. In response to public pressure, the Ecuadorean National Forestry, Natural Areas and Wildlife Institute, INEFAN, initially denied, then granted, and subsequently revoked permission to Petroecuador to work in pristine areas of Cuyabeno on the grounds that Petroecuador was not complying with environmental laws.

The Huaorani are also uniting to protect their territory. After a failed effort to prevent Maxus from developing oilfields in Huaorani territory, the organization, ONHAE, signed a twenty-year friendship agreement with the company in 1993. However there is a tremendous gap between what it means to them and what it means to the company. For example, one woman who attended the signing ceremony described it as an agreement for t-shirts. In 1995, the leadership of ONHAE changed. The new leadership is more critical of Maxus, and has threatened to take direct action against not only Maxus but also Oryx, Elf Aquitaine and Petroecuador, in order to force all of the companies working in their territory to negotiate with the Huaorani, honour their commitments, and protect the environment.

In another promising initiative, the indigenous umbrella organization, Confederation of Indigenous Nationalities of the Ecuadorean Amazon, CONFENIAE, is developing a regional socio-environmental monitoring programme with the Quito-based environmental group, Acción Ecológica. To date, however, the effectiveness of the project has been seriously hampered by limited financial resources, and a lack of technical expertise, access to information, and cooperation from the government and industry.

Many environmental groups in Ecuador, such as the wealthy and powerful Fundación Natura, have shied away from publicly criticizing the oil industry, especially foreign companies. Acción Ecológica, on the other hand, has played an important role in garnering international support for a boycott of Texaco (launched in Norway in 1992 and in the US in 1993). Through the monitoring programme, it is also beginning to work with some local groups in the field. In addition to CONFENIAE, Acción Ecológica works with the Front for the Defence of the Amazon ('the Frente'), founded by colonist organizations around Lago Agrio in response to the lawsuit against Texaco in New York. Grassroots concern about Texaco's legacy is the common ground that is fuelling increased activism and alliances between growing numbers of colonists from Ecuador's coastal and highland regions and different groups of indigenous peoples. The Frente is reaching out to indigenous organizations such as CONFENIAE and to colonists elsewhere. One important group that has joined the Frente is the Federation of Campesino Organizations of Napo (FOCAN-Napo) in Coca. FOCAN-Napo was founded by colonists who were among the first Oriente residents to denounce contamination in the area. Ecuador's Amazon is at a crossroads.

The future of the region's natural resources and its peoples is seriously threatened by current development policies. Oil production facilities are now being operated by Petroecuador, Clyde Petroleum, Occidental Petroleum, Oryx Energy, Elf Aquitaine, and Maxus Energy Corp. (recently purchased by Yacimientos Petrolíferos Fiscales, YPF). ARCO has completed exploration and plans to begin production after the capacity of the trans-Ecuadorean pipeline has been increased. Exploratory activities are beginning or expected in new concessions operated by Oryx, Santa Fe Energy, Amoco/Mobil, Tripetrol, Triton, and City Investing, and in existing concessions operated by Occidental and Petrobras. The impacts that have occurred to date could just be the tip of the iceberg.

Conclusion

The jury is still out on whether any company can extract oil from a sensitive rainforest environment – using current technology – without serious injury to the land and the people. Good faith efforts to 'minimize' environmental impacts may not be adequate to maintain ecological integrity, and protect natural resources that are vital to the health and well-being of Amazonian peoples and sustainable development in the region. Moreover, many companies – including some US, European, Latin American, and Asian operators, investors and contractors – prefer to maintain a low profile, and, to date, their operations have not differed significantly from Texaco and Petroecuador. In addition, after a lull in the early 1990s in the wake of international criticism of environmental conditions in the oilfields, public international institutions such as the World Bank and the European Union are expected to finance the government and even Petroecuador, for oilfield activities. Funding proposals carry environmental labels; however they were not developed in consulation with affected populations, and have not been accompanied by adequate mechanisms for public disclosure and independent monitoring.

Without transparency and independent monitoring, residents, environmentalists, and policy-makers cannot know the true environmental and social costs of oil development. In addition to supporting Oriente peoples in their demands for participation and accountability from public institutions and transnational corporations, the industrial world has a responsibility to take steps to better understand the links between consumption in the North and environmental destruction and human rights violations in the South. Because the encroaching industrial world wields overwhelming technological, political and economic power, Oriente peoples are increasingly looking beyond their worlds, to build local, national and international alliances. To varying degrees, these alliances are gathering strength, but they remain fragile. To succeed, they will need international political, technical and economic support, accompanied by a genuine respect for the ethnic, territorial and environmental rights of Oriente peoples.

Suggested Reading

Laura Chinchilla and David Schodt, *The Administration of Justice in Ecuador*, Center for the Administration of Justice, Florida State University, 1993.

David Corkill and David Cubitt, *Ecuador: Fragile Democracy*, Latin America Bureau, London 1988.

Embassy of Ecuador, 'Diplomatic Protest signed by Edgar Teran, Ambassador of Ecuador to the US, to the US Department of State', No. 4-2-138/93, Washington, D.C., Dec. 3 1993, Unofficial Translation, in Appendix to Texaco Inc.'s Motions to Dismiss, Aguinda v. Texaco, No. 93 CIV 7427 (S.D.N.Y.) (Dec. 28, 1993).

Joe Kane, 'Letter from the Amazon: With Spears From All Sides' New Yorker, Sept. 27, 1993

Judith Kimerling, *Amazon Crude*, NRDC, 1991; 'Rights, Responsibilities and Realities: Environmental Protection Law in Ecuador's Amazon Oil Fields', forthcoming in *Southwestern Journal of Law and Trade in the Americas*; 'Dislocation, Evangelization and Contamination: Amazon Crude and the Huaorani People,' in *Ethnic Conflict and Governance in Comparative Perspective*, Working Paper Series No. 215, Woodrow Wilson International Center for Scholars, 1995; 'Disregarding Environmental Law: Petroleum Development in Protected Natural Areas and Indigenous Homelands in the Ecuadorean Amazon', *Hastings International and Comparative Law Review* 14: pp 840-903, 1991.

Colombia's Pacific Plan: Indigenous and Afro-Colombian communities challenge the developers

Lucy Alexander

> The land doesn't belong to one person, but to everyone. We say that 'man is of the land, but the land is not man's'. So we say that everything that happens to the land will happen to its children. And that doesn't just mean indigenous people, but you here.
>
> Euclides Peña Ismare, Emberá-Wounaan indigenous leader, May 1995

The Pacific coast region of Colombia is a little known arena for the fight between local communities and developers in Latin America. For the past few years, the Afro-Colombians and indigenous peoples who have occupied this area of tropical rainforest for centuries have been struggling against the Colombian government to defend their land and their livelihoods against plans for intensive development.

The basin of land between the western-most range of the Andes and the Pacific coast covers over six per cent of Colombian territory. It is a vast region of tropical rainforests, mangrove swamps, cloud and montane forest. Until recently, two-thirds of its 7.3 million hectares of rainforest were primary forest. Scientists consider it one of the world's most biodiverse areas; there are more species of tree in just one hectare of the Pacific's rainforest than in the whole of the British Isles. There is also a particularly high proportion (about 25 per cent) of endemic species. Rivers and their tributaries criss-cross the region, which has one of the world's highest annual rainfalls. The majority of the rural communities are riverine, either living from subsistence farming and fishing if they are Afro-Colombian, or hunting and gathering if they are indigenous.

The Pacific region also possesses immense natural resources. There are abundant mineral riches of gold, platinum and oil lying underground. Above ground, thousands of hectares of prime forest provide much of Colombia's timber, and large rivers can be harnessed for hydro-electric power. Mining has been carried out in the region for at least three hundred years. The mines were originally operated by African slaves brought over to replace the local indigenous peoples who refused to take minerals out of the earth. Geologists believe, however, that the mineral wealth of the Pacific region is still largely untapped. It is not for nothing that successive Colombian leaders in recent years have been looking for ways to exploit what they call their country's 'money-box'.

For at least three hundred years, Afro-Colombians and indigenous peoples have lived side by side in the Pacific, or the Chocó as it is known after the name of its largest department. Of the 817,000 inhabitants of the region, about ninety per cent are Afro-Colombians and the rest of the population is divided almost equally between mestizo (mixed race) and indigenous peoples. There are five distinct Indian groups: Emberá, Wounaan, Tule, Eperara Siapidara and Awa. The Pacific basin has one of the highest concentrations of indigenous peoples in Colombia, which nationally has a total of 87 different indigenous groups. Colombians link the Indians with a proud defiance of colonization going back to the fifteenth century. This is particularly true of the Pacific, which historians believe was penetrated by fortune-seekers very soon after 1492 and has been looted and exploited by outsiders ever since.[1]

The Chocó's inhabitants

This history will be significant when we look at how indigenous peoples are facing the new wave of development currently taking place in the Pacific. Since the Indians of the Chocó have defended their land and culture from outsiders for centuries, the new threat simply fits into this pattern of external forces trying to plunder their natural resources and the land they occupy. Speculators have always come to the area for mining, logging, or ranching. Even today, when intensive and technologically sophisticated mining has still not reached the region, it is Colombia's main producer of platinum and second producer of gold because deposits are so large.

But as with so many parts of the world rich in natural resources, the local inhabitants of the Colombian Pacific basin are not the main beneficiaries of the wealth generated by this area. Colombian statistics show that over 80 per cent of the inhabitants of the Chocó do not have basic necessities like an adequate supply of potable water or sanitation. Per capita annual income is the lowest in Colombia and among the lowest in Latin America at US$500. Over half the population lives in absolute poverty. Added to this there is an infant mortality rate of 150 per 1,000 live births and life expectancy is only 55. This poverty means that there has been a tradition of emigration from the Pacific, particularly by Afro-Colombians, who drift to the cities to work as street-sellers, in domestic service or to study.

Many of the Chocó's inhabitants, whether black or indigenous, have remained rural and riverine, however. They depend on the rivers for food, travel and sanitation. Yet in areas where the extensions of the Panamerican Highway are being built, whole stretches of river have been filled with mud and sawdust, forcing local inhabitants to move. Elsewhere, rivers have been widened and their courses changed to allow for the transport of logs. One resident from the River Patia area had to rebuild his house five times as the river was widened to feed the nearby sawmills. 'You used to be able to throw a guava across the Sanquianga, now it's thirty times as broad and we've lost most of our farmland,'says Georgia Castro.[2]

The Plan Pacífico

The lynchpin of the region's current development onslaught is the so-called 'Plan Pacífico', officially launched in 1987 by the then president Virgilio Barco Vargas. But the dream of reaping huge benefits from the country's 'money-box' has been teasing Colombian leaders for decades. In 1973 the British explorer Robin Hanbury-Tenison was invited to assess the impact of extending the Panamerican Highway across the Darien Gap. He concluded that the road could do nothing but harm to the local communities:

> Their situation is particularly vulnerable because they do not benefit from the short-term gains of exploiting the jungle with which they have always lived in harmony. They have excellent access and transportation through the network of rivers, so that the road will not even represent a potential economic outlet for them. Nor do they regard the jungle as a threat. Indeed, if it is removed they are likely to die out.[3]

These sentiments and the feelings of local communities in the Chocó have not deterred the planners. With its extensive Pacific coast, presidents as far back as 1973 foresaw a profitable expansion into the Pacific rim, enhancing Colombia's already advantageous position as an axis between North, Central and South America. The region has sometimes been portrayed as Colombia's last frontier; a wilderness of rain and forest and rivers, cut off from the rest of the country by a spur of the Andes and by the wretched living conditions of most of its inhabitants.

However, pressing economic reasons have added a new urgency to the plans to develop the Pacific. Neo-liberalism, which has swept through most of Latin America in the past decade, reached Colombia relatively late. Its advent, along with an International Monetary Fund structural adjustment package of economic cutbacks, has led to the usual string of privatizations, new government institutions and social spending cuts. It has also meant that Colombia, as part of its structural adjustment programme, must urgently generate revenue to pay off its external debt of about US$21.5 billion.[4] Exploitation of the Chocó is now considered essential for the economy to grow under its free-market or neo-liberal model.

The export potential of the region is vast and is seen as a key to the markets opening up all over Latin America. Further impetus has been added to government interest in rapid development by the Trade Agreement which Colombia has entered into with Mexico and Venezuela. Colombia hopes to gain access to NAFTA via this 'Group of Three', and so must keep up its potential for trade with these countries. It was in this context that President César Gaviria launched yet another plan, grandly called 'The Pacific Plan: A New Strategy of Sustainable Development for the Colombian Pacific coast' in 1992. The wording of this new document was subtle and careful, avoiding terminology resonant of the 'megaprojects' specified by his predecessor Barco in his plan of several years earlier. However, its overtones of care for the environment and inhabitants of the Pacific have not allayed the fears of the latter.

Buried in the middle of the new 'Plan Pacífico' paper, after sections with reassuring titles like 'Health', 'Basic Sanitation' and 'Protection of Biodiversity', comes the real agenda for developing the region. Under the new Plan, the hydroelectric dams, roads, railways, canals, naval bases, docks, land bridges, oil wells, mines, telecommunications, fishing and forestry projects which have long been on the agenda for extracting maximum revenue from the Pacific are still being budgeted for. In fact, considering the relatively slim budget of US$321 for the 1992-94 expenditure of the entire Plan Pacífico, it is difficult to imagine that everything listed in the Plan could be paid for. Social projects, which are given such prominence in the document, are not supported by a similarly generous allowance in the budget. The lion's share of cash is going to the same 'megaprojects' outlined by Barco.

The inevitable consequences of the megaprojects – environmental damage and social disruption – have already begun. The projects include the hydroelectric power station at Urra, which has affected Emberá indigenous territory around the Esmeralda and Verde rivers and perforations for petroleum, gold-mining and road construction. Already, many Pacific peoples have lost their homes, their rivers and their livelihoods. The prospects for the Pacific region's flora and fauna are not good if the bulk of the development described in the Plan Pacífico takes place. One of the best-known aspects of the Plan is the construction of new roads, branches and extensions of the Panamerican Highway. This highway, which already runs down the west side of the Americas, from Alaska to Southern Chile, is broken for just 54 kilometres in the area around the Darien Gap. This land of swamps and cloud-forest, like much of the Colombian Pacific region, has its own endemic species, including Goldman's Humming-bird and Quail Dove, Golden Headed Quetzals, the Pirre Warbler and the Panama Tapaculo. These exotic tropical birds would lose their natural habitat if the road were built. The Atrato Swamp region is home to a similarly long roll-call of species, also threatened by the roadbuilding. In all, up to 10,000 acres of wilderness will be destroyed if all the extensions of the Panamerican Highway are completed.[5]

Pacific people's responses

The people of the Pacific have many responses to what is happening to their habitat, and over the past decade have been consolidating their reactions and defending their rights. Some say they are not necessarily opposed to development, but they fear their interests will never be upheld by the government's development schemes, despite the caring wording of all the recent documents. This mistrust has not been erased by the newest government proposal, the 'Biopacífico' project, launched in March 1993 with funding from the Global Environmental Facility.

Under the 'Plan Biopacífico', which is being run by the new Ministry for the Environment, the government has named one of its most important tasks as getting local communities on their side. It is also emphasizing the importance of the Pacific's bio-diversity, as its name would suggest. To do this, the government proposes to make a register of all species of flora and fauna in the area in order to chart their destruction or survival. Yet again, however, the government has alienated the inhabitants of the Chocó. The principal indigenous organization in the Chocó, OREWA, says there is no need for a register, because the people themselves know what plants and animals exist and will notice if they disappear. They are concerned that the state will try and claim intellectual ownership of their biological and botanical knowledge of the Pacific. They see this as another potential rape of their resources, following extraction of the gold, platinum and petrol from their territory.[6] For hundreds of years, the local inhabitants of the Pacific, particularly the indigenous peoples, have used their knowledge of the properties of plants and animals in traditional medicines. Now they fear that a formal register of this information will be sold to pharmaceutical companies by the government.

Afro-Colombians and indigenous people are wary of any government moves to develop their homeland, whatever the name of the plan it tries to do this under, but they are not opposed to development *per se*. Living in an area with a wealth of natural resources, they are aware that they have been marginalized by Colombian governments for too long. This means that they look for ulterior motives in the Plan Pacífico and the Plan Biopacífico but they are also concerned with setting an agenda for developing the region that is their own. Since the mid-eighties, a plethora of community movements has sprung up all over the Pacific basin, representing the Afro-Colombian and the indigenous people. In many cases, the two communities began by operating independently, but recently there have been signs of a new phase of consensus between these sometime opponents.

One of the first major victories for the new civic movements against the Plan Pacífico is the stalling of one of the major new roads. In July 1992, about four hundred indigenous people from communities which will be affected by the Pereira-Nuquí branch of the Panamerican Highway staged a sit-in at the diggers' site. They stayed for two weeks, stopping the building immediately and did not leave until they had received assurances that a study would be made of the environmental impact of the road. This study has recently been finished and the results have not yet been released. However, indigenous representatives have been sceptical since the outset that the study will endorse plans to build the road.

Such worries are confirmed by recent news that the plan to build an international port in Tribugá Bay, near Nuquí, is going ahead.

The indigenous communities plan to continue to press for their rights to be respected. Their main demands are 1) that formal guarantees are made to ensure that colonization and exploitation will not take place around the road, 2) that the government undertakes a programme to improve basic services in the region, especially health and education and 3) that indigenous and black land rights in the area around the road are respected. Land rights are fundamental for the local communities in the Pacific. Some Indians feel that their territorial claims are stronger than those of the Afro-Colombians. Today, although only 4 per cent of the inhabitants in the Pacific are indigenous, they hold 26 per cent of the land.

Indigenous groups have been articulate and organized about defending their land rights and the government, foreseeing trouble over land in the future, included clauses on indigenous land rights and cultural sovereignty in the new constitution. 'Transitory Article 55', the clause which defined indigenous land rights, was perceived as divisive by the communities of the Pacific, because it did not automatically grant Afro-Colombians the same entitlement to their land. This might have been due in part to the presence (by the time the new constitution was drawn up) of indigenous politicians in Congress. 'Transitory Article 55' granted Afro-Colombians rights only to what it termed 'vacant lands', an expression which made the Afro-Colombians feel that their historical links with the land they occupied were being denied and at the same time, in being labelled 'vacant', that it was going to be offered to outside developers if they did not gain titles quickly.

One of the most dynamic forces in championing indigenous rights is OREWA, the Regional Indigenous Organization of the Emberá Wounaan. OREWA, along with many other community movements, does not oppose all development in the Chocó. With groups like the Antioquia Indigenous Organization (OIA) and the Organization of People's Neighbourhoods of the Chocó (OBAPO), it is pressing for indigenous and black peoples to be included in the development plans and to be called upon for their opinions.

OREWA is now leading the new rapprochement between Afro-Colombians and Indians, which it began by helping push through Law 70. This law, passed in 1993 after being introduced by an indigenous member of congress, has granted land titles to Afro-Colombians for the first time. Unfortunately, in the short term this is causing problems over titling, as some violent, even murderous, conflicts have taken place when Afro-Colombians and indigenous peoples have fought for the same plots of fertile land. These plots can be scarce, as much of the land near the inhabited areas of the Pacific is already exhausted from subsistence farming or logging.

Historically, there has been a divide between black and indigenous peoples of the Chocó, but, on the whole, the two communities have avoided conflict. The Plan Pacífico has stirred up the differences between these communities, but it is also beginning to have some unexpected benefits. In radicalising black and indigenous communities, it has made each group more united and articulate about defending their rights. While the Indians have been organized for decades, this has had a great impact on the Afro-Colombians, who have not had such a strong history of defending their rights and have often been perceived as victims of capitalist exploitation in Colombia. Now, several Afro-Colombian organizations exist and are looking at ways to stop their people losing out under the Plan Pacífico. 'My enemy's enemy is my friend' is perhaps one way of describing what is beginning to happen in the Colombian Pacific. OREWA is pioneering work to create inter-ethnic harmony. With funding from the British overseas charity, Christian Aid, it will run workshops for black and indigenous leaders to examine how different aspects of the Plan Pacífico will affect their people, focusing on land rights and how this new external force has brought conflict to communities which have lived side by side for generations. The Indigenous Pastoral Centre is running a scheme to promote racial integration by placing Afro-Colombian families in remote indigenous communities.

This kind of cooperation is vital if the people of the Pacific are to preserve their land and have a voice in how it is developed. The indigenous leader, Euclides Peña Ismare, says 'For us, the forest is heaven. The life of our people is heaven, as they live in nature, we don't have another heaven to go and seek'. It would be another environmental and human tragedy if this heaven were destroyed. It would be a human victory if two historically opposed communities managed to work together to make the development take place only on their terms.

[1] Jon Barnes, 'The Colombian Plan Pacifico: Sustaining the Unsustainable', CIIR Occasional Paper 1993.

[2] Sarita Kendall, 'Conservation by Consensus in Colombia's Pacific Frontier,' *Choices Magazine,* May, 1995.

[3] 'Should the Darien Gap be closed?'. A.R. Hanbury-Tenison, *Geographical Journal*, Vol 139, Part 1, Feb 1973

[4] Duncan Green, *Silent Revolution – the rise of market economics in Latin America*, LAB/ Cassell, 1995.

[5] 'Plan Pacifico: a special CUSO report.' CUSO, Ontario, Canada, Nov 1994.

[6] Sarita Kendall, op cit.

3
FIGHT FOR THE FOREST

Amazonian Indians and peasants: coping in the age of development

Stephen Nugent

Brazilian Amazonia has been an arena of both intense social change and intense scrutiny in recent years. Following the construction of the Transamazonica Highway in 1970, the region has come to typify a process of modernization both villified and celebrated. It has also become the subject of much social and natural science research as well as journalistic speculation about what this so-called 'green hell' will look like once the scaffolding is down. From the perspective of the Brazilian state, Amazonia has traditionally been of marginal concern. Little involved in the calculations which rate Brazil as a major economic power, Amazonia has been depicted officially as a stagnant backwater of high potential where the forces of nature have thus far prevailed over the best intentions of both its occupants and its would-be modernizers. From the vantage points of Europe and North America, Amazonia has been a folkloric repository of exotic peoples, plants and animals, a primeval space in which the archetypes of the tropical New World still obtain.

Transamazonica and the Brazilian government's Plan of National Integration may have changed this perception but two images persist. One of these is Amazonia as a natural landscape in which society is contingent, tolerated, and constantly deferring to the rigid conditions of the humid tropical forest ecosystem. The second is Amazonia as a particularly fraught social landscape: land conflict, predatory goldminers, persecuted Indians, urban squalor. While both images reflect undeniable realities and are the starting points for much excellent critical commentary on recent Amazonian affairs, they also feed a vision of a postponed Amazonia, one in which attention to the current, frequently tumultuous and confused, situation is passed over in favour of consideration of possible – dire or utopian – outcomes. Overlooked are Amazonian societies (both indigenous and postcolonial) which, despite the uncertain conditions in which they carry on, not only display possibilities for a non-pathological Amazonia, but are already concrete and integral.

Documentation of social conflict and environmental depredation in Amazonia is extensive and growing. It has been spurred by compelling arguments to the effect that unregulated development in Amazonia threatens to produce irrevocable – and largely negative – changes in the structure of the humid tropical forest(s) and the lives of those who occupy them. There are few grounds for balancing this critical consensus with an up-beat presentation of the virtues of such development. But the terms of debate which prevail on the future of Amazonia exclude a crucial perspective, that of Amazonians who are not registered participants in the debates, but who cope on the ground, as a daily matter of course, with the very forces which have prompted the wider, globalist concern about the 'future of Amazonia'. These are peoples named from afar as 'Indians' (e.g. the Kayapo, the Wai-Wai), as 'peasants' (e.g. *caboclo*, *campesinos*), as colonists (e.g. 'spontaneous' or 'planned', or Japanese), not infrequently simply in terms of their imputed roles (e.g. frontier folk, pioneers, smallholders, fisher folk, entrepreneurs, etc.) and often as mere 'populations' or 'inhabitants'.

Although represented nationally and internationally through various nominally progressive fora (academic, human rights, environmentalist), Amazonians generally evade clear definition. Consequently, how they conduct their lives while globalist debate continues is rarely depicted. What I intend to illustrate here is that within this precarious social landscape, the conditions of existence for Amazonians are not driven exclusively by the protocols of globalist debates about the future of Amazonia. Although such debates play a vital role, they tend to assume that the future of Amazonia depends on analyses and solutions generated by privileged – and generally remote – overseers. There is no question that the critical attention paid the plight of modern Amazonia by external, interested parties has enhanced the prospects for a just resolution of the problems facing Amazonia and Amazonians, but such support should not obscure what Amazonians themselves are doing and have done.

The doctrine of 'tropical nastiness'

The view that Amazonia is an economic backwater is closely related to what the geographer J.M. Blaut has referred to as 'the doctrine of tropical nastiness'[1] and which asserts that tropical soils are a limiting factor in social development. There is general agreement that Amazonian soils are largely inappropriate for the agriculture and pastoral use of temperate climates. Yet even this view can be refuted by the experience of Japanese colonists in Amazonia (particularly in the settlement of Tomé-Açu, in the state of Pará). Within thirty years of its founding (c. 1929), Tomé-Açu had become the first significant source of black pepper in the western hemisphere, accounting for more than five per cent of world production. The development of such an industry is remarkable in itself, but from the perspective of the tropical nastiness doctrine, even more remarkable is the fact that such agriculture has taken place on permanent sites (some of which have been under continuous cultivation for fifteen years) and that the average cultivated area of Japanese farms (twenty ha.) is well below the typical area for Amazonian smallholders (whose agriculture is regarded as predatory and unsustainable). Black pepper provides a standard of living unusually high for the region and accounts for almost 80 per cent of the income of those who produce it; it is just one of an extensive repertoire of marketable crops integrated into a form of agro-forestry (whereby crops are grown in harmony with the forest's trees). Although black pepper requires significant inputs of fertilizers and pesticides, it is clear that assumptions about the intrinsic obstacles to agricultural development in the humid neo-tropics demand revision.

The success of Japanese colonists in Amazonia has attracted much less attention than one might imagine, reflecting the durability of the tropical nastiness doctrine. Some claim that Japanese success in Amazonia is 'interesting' but 'anomalous'. Subler and Uhl assert that what is crucially different about the Japanese of Tomé-Açu is that they have security of land tenure and access to agricultural credit, conditions still beyond the grasp of many Amazonians.[2]

Forest managers

In 1989 the protest at Altamira (a major town on the Transamazonica Highway) against the construction of a hydroelectric dam complex which would have flooded vast areas of forest occupied by various Amerindians (and neo-Amazonians), drew particular attention to the Kayapo Indians, among the groups most directly affected by the proposed dams. The Kayapo are perhaps the most visible of Gê-Bororo peoples (a linguistic designation) who occupy the savannah region of central-southern Amazonia and whose livelihoods have been the subject of intense dispute among anthropologists and others.[3] Debate centres around the extent to which native peoples' long-standing manipulation of 'green hell' offers possibilities for mitigating the effects of rampant developmentalism.

Darrell Posey and his supporters have argued that the forest islands (*apêtes*) occupied by the Kayapo which dominate the region are the products of human intervention: man-made forests designed to meet the immediate and long-term needs of the Kayapo. Critics such as Parker, on the other hand, have disputed the human origins of such islands, claiming that the composition of apêtes differs insignificantly from that of naturally-occurring islands. Aside from the specifics of the debate, the notion that native resource-management can be exploited in diverse ways is at least firmly on the agenda.

Those who believe that traditional, indigenous ways of life ought to be maintained argue that: 1) their accommodating strategies vis-á-vis the humid tropics are viable;[4] 2) they have constructed environments appropriate to the needs of local peoples. Far from being passive actors constrained by the rigidities of the tropics, it is argued that peoples such as the Kayapo have manipulated their environment for the good of their societies.[5] 3) the 'wise forest-management' of indigenous communities offers potential for 'ethical commerce'. Through collaboration with the commercial world, it is claimed that native peoples can market traditional products and use the proceeds to maintain their integrity as autonomous 'wise forest-managers'.[6] The Kayapo, for example, produce Brazil nut oil on behalf of Body Shop International (BSI) which is used to manufacture hair conditioner, a leading Body Shop product.

It is possible that the use of 'green' tropical forest products by BSI or in Ben and Jerry's Rainforest Crunch ice cream may improve the lot of the Amazonian producers. But whether such forms of commercial marketing provide the basis for major revisions of development policy remains an open question. Critics of the Body Shop rationale[7] argue that the success of such arrangements is open to speculation and that the beneficiaries bear unequal risks. For BSI, the failure of such schemes is unlikely to be life-threatening. For the Kayapo (and other native peoples drawn into the world of fair trade), the risks are rather greater. The commercialization of the products of 'wise forest-management' may help mitigate the effects of native peoples' integration into the national and global economy.

Arguments about how to maintain native peoples' autonomy in the face of developmental onslaughts have not yet been resolved. But generally there is at least a measure of support for the idea that people living in Amazonia actually know something about what they are doing. While they may have been vague about how their proposals were to be implemented, Peters, Gentry and Mendelsohn have carved out a space in which both native peoples, colonizing peasants and remote investors could have their cakes and eat them.[8] In their view, the extraction of non-timber forest products (NTFPs) could be more profitable than the major uses of neo-tropical forests – cattle ranching and logging – and would avoid untoward environmental effects. The viability of such efforts is likely to be judged on the basis of a small number of projects (e.g. Body Shop and Ben and Jerry's; most of those marketing tropical products are less visible) and attention will be focused on those seen to be delivering economic results (i.e. the retailers). But there are *local* success stories with non-timber forest products[9] which support the viability of such alternative forms of forest use, albeit on a reduced scale.

The acceptable face of cattle ranching?

The two examples I have just presented have dealt with Amazonians directly involved in the global economy. The third example does not. A major feature of Amazonia has been deforestation. Some view deforestation as an inevitable and necessary consequence of development, and others view it as the number one reason for stopping further development in Amazonia. Then there is another group of people who have little choice in the matter: deforestation is an established feature of the social and natural environment they occupy.

It is quite clear that the causes of deforestation are not intrinsically 'Amazonian'. At the time of conquest, most Amazonian peoples occupied riverine rather than forest areas and the region supported – without negative environmental effects – a population comparable to that of the present day. Deforestation has been the result of an external, national and international, demand for cheap resources. For example, most of the twelve per cent of Amazonia which is now deforested was cleared because the Brazilian government encouraged people to do so (either through financial incentives or through road-building schemes). By contrast, the terms of reference and markets for local producers are either local or regional.

Despite the negative consequences of deforestation at a macro level (loss of plant and animal biomass, species extinction, soil hardening, silting up of water courses, loss of the forest's capacity to absorb and recycle carbon waste products), for many Amazonians these are not merely adverse environmental changes, but are their conditions of existence. Eastern Pará is an Amazonian region seriously affected by forest clearances in the name of timber and cattle pasture. It was here in the early 1990s that Mattos and Uhl[10] analysed what happens to those who remain on degraded land. Their findings appear to contradict the commonly held view that such people are an ungainly, predatory peasantry. In their survey of pastoral enterprises ranging from <100 ha. to >500 ha., they discovered that large ranches are less profitable than small ones. This was not surprising, given that much of the profitability of large-scale ranches is derived from subsidies and tax breaks. But they also found that, even in 'green hell', cattle ranching may be viable. In view of the widespread condemnation of cattle ranching, this finding *was* surprising.

Mattos and Uhl produced uncomfortable results for two opposed camps. From the point of view of those who have promoted large-scale cattle-ranching as an appropriate use of cleared neo-tropical forest (for which there is little, if any, supporting evidence), the Mattos and Uhl study offers little comfort. From the point of view of those who regard Amazonian land as suitable only for 'traditional' pursuits, the results are also uncomfortable, for they reveal that Amazonians have their own agendas, one of which includes cattle ranching, albeit on a small scale and for local, rather than export, markets. The study shows that intensive/semi-intensive dairy (and to a lesser extent beef) farming on recovered forest soils can provide a relatively high standard of living and through careful management (including significant investment of capital), such farming is sustainable in both environmental and social terms. Lest it be thought that such research overturns widely-held views about the inadvisability of cattle ranching in the Amazon, it is important to note the small-scale ranchers studied by Mattos and Uhl are not speculative entrepreneurs for whom Amazonian land provides a cheap resource,[11] but are small landholders integrated into local markets and are producers for whom cattle ranching is one of a repertoire of agricultural pursuits. (By contrast, of the large-scale ranchers, fifty-nine per cent began ranching as soon as they arrived in the region and ninety-three per cent of such ranchers were also involved in logging activities.)[12]

The dominant image of Amazonian cattle ranching has been one in which 'the hamburger connection' features prominently and as the authors note,

> it is hard to justify using Amazonia to raise meat for other regions of Brazil given the high ecological costs involved in Amazonian ranching and considering that other regions of Brazil are much better suited to ranching in terms of climate, soils and infrastructure than Amazonia[13]

But they also observe that 'the debate is no longer over whether cattle belong in the Amazon. Ranching is here to stay.'[14]

While the research agenda in Amazonia has been driven by the search for forms of appropriate or sustainable development, increasingly, it has been found that many desirable practices are already in place.

Caboclos and Neo-Amazonians

The emergence in the 1980s of the Rubber Tappers Union as a force in Amazonian politics was not without its ironies, tragedies and maladies (Chico Mendes's representation as a tree-hugger rather than socialist activist; Chico Mendes's assassination; the continued persecution of activists), but while the rubber tappers (and other groups) enlarged the political space within which Amazonia's grassroots organizations were seen to (and do) operate, the characterization of other Amazonians was often consigned to stereotype. There are fifteen to eighteen million Amazonians, and relatively few are rubber tappers or Amerindians (Allegretti estimates that 68,000 families are dependent on rubber extraction[15]). The remaining Amazonians are heterodox – farmers, taxi-drivers, 'wise forest managers', chancers, ambitious schoolchildren, and on and on – a fact unsettling for those who prefer to view the social landscape of Amazonia as undeveloped and still to be mapped. Despite such prescriptions, Amazonians – like other stereotypical others – do and do not adhere to the roles allocated them by their observers.

The history of Amazonia is largely told from the perspective of conquerers and other interlopers. What has happened in the region during times of declining economic activity is relatively poorly documented. The two recent, major episodes of intense outsider interest in the region were the rubber period (1820-1910) and the period commencing with the construction of the Transamazonica highway c.1970. In the intervening period, although there was a reduction in economic export activity, it is hardly the case that Amazonian society went into decline. Indeed, it could be argued that during this period, one of the most plausible development models emerged in the form of '*caboclo* society'.

The term *caboclo* is generally pejorative and in Amazonia it refers to people whose identity is more regional than national (a Brazilian version of 'hill-billy' or 'country bumpkin'). When it comes to the 'tropical nastiness' doctrine, the caboclos bear the brunt of the defects attributed to the neo-tropics; but lurking beneath this negative characterization is something else. With the collapse of the rubber industry (c.1910), Amazonia was virtually relinquished to caboclos. With the rediscovery of Amazonia at the time of the implementation of the Plan of National Integration in the early 1970s, the fact that caboclos had been meeting the challenge of the tropics for half a century was hardly voiced. Instead, Amazonia was again treated as a vast, untapped resource, development of which could only be achieved by the engagement of technocrats, bureaucrats, and consultants (few of whom would defend their achievements today).

Because Amazonian societies were regarded as being dependent on outside forces for their dynamism, those already living in the region were largely ignored when Amazonia was rediscovered. They were described simply as 'populations' or 'inhabitants'. Yet many Amazonian societies neither disappeared after the rubber boom nor experienced much change following the construction of the Transamazonica highway. The historical peasantries of Brazilian Amazonia (here glossed as caboclos) significantly diversified their livelihoods in the post-rubber period while consolidating their positions in the most desirable niches of Amazonia. As with pre-colonial, native societies, caboclo societies favoured riverine locations with access to the fertile *várzeas* replenished by the annual flooding of the river. While they do have specialist occupations, most caboclos are engaged in a range of livelihoods. Given that caboclo societies emerged from within the devolved apparatus of the rubber industry, they were already integrated into regional, national and international trade, albeit on a much more precarious basis than during the rubber period. Jute, for example, became a staple product after the 1920s and small-scale cattle-rearing became an established feature.

Growth and integration of local and regional markets is conventionally overlooked in contemporary accounts not least because the effects of Transamazonica-related development were so abrupt and so severe. Two examples illustrate aspects of caboclo dynamism effaced by recent developments. One of these is the emergence in the estuary of the Amazon (the confluence of the Guamá, Amazon and Tocantins rivers) of cane rum (*cachaça*) production for local markets and for export upriver. These involve cultivation and processing of sugar-cane, on-site distilling and bottling and – significantly – the use of tidal energy.[16] With Trans-amazonica appeared mass-produced brands which quickly drove local products from the market, but until that point what was evident was a level of commodity production far removed from the received view of caboclos as marginal forest hunter/gatherer/horticulturalists. The second example is provided by Toby McGrath's study of long-distance trading boats (*regatões*) which ascend the river from the estuary bearing manufactured goods exchanged for the forest/riverine products of remote communities (as far away as Tefe, above Manaus).[17] Such networks of trade show not only the durability of inter- and intra-regional production and trade links, but also the relative invisibility of such links: this trade takes place beyond the gaze of official agencies and strictures of the formal market.

Conclusion

The resilience of Amazonian societies in the face of pressure from top-down developmentalism is increasingly acknowledged in recent literature, but there is still a tendency to treat such examples of successful accommodation as extraordinary rather than typical. In part, this tendency reflects the overall research agenda which has focused on how wealth can be extracted from Amazonia, but it also reflects a seemingly intransigent commitment to the idea that the solutions to the so-called 'dilemmas of Amazonian development' will come from the intervention of appropriately armed external allies. While not wishing to discount such sympathetic interventions, it should be noted that so far the best examples of what the future of Amazonian societies may hold have come from actual societies in Amazonia, not from the drawing boards of remote supervisors.

[1] J.M. Blaut, *The Colonizer's Model of the World: Geographical Diffusionism and Eurocentric History*, Guildford Press, New York, 1993.
[2] S. Subler and C. Uhl, 'Japanese agroforestry in Amazonia: a case-study in Tomé-Açu' in A. Anderson (ed), *Alternatives to Deforestation: Steps towards sustainable Use of the Amazon Rain Forest*, Columbia University Press, New York, 1990.
[3] See E. Parker, 'Forest islands and Kayapo resource management in Amazonia: a reappraisal of the apete' in *American Anthropologist* 94:2, pp 406-28; D.A. Posey, ' Reply to Parker' in *American Anthropologist* 94:2, pp 441-43; see also A. Anderson op cit for wider discussion.
[4] D. Irvine, 'Succession management and rainforest distribution in an Amazonian rainforest' in D. Posey and W. Balée (eds), *Resource Management in Amazonia: indigenous and folk strategies (Advances in Economic Botany Vol. 7)*, New York Botanical Society, 1989.
[5] D. Posey and W. Balée op cit 1989.
[6] G. Roddick, 'Reply to Corry', *The Ecologist* 23: 4:pp 198-200, 1993; J. Clay, 'Report on funding and investment opportunities for income generating activities that could complement strategies to halt environmental degradation in the greater Amazon basin', unpublished ms., n.d. 1988 and *Indigenous Peoples and Tropical Forests,* Cultural Survival, Cambridge, MA, 1992.
[7] S. Corry, 'The Rainforest Harvest: who reaps the benefit?' *The Ecologist* 23:4: pp.148-53.
[8] L. Peters, A Gentry, and Mendelsohn, 'Valuation of an Amazon Forest', *Nature*, 339:655-6
[9] See Scott Anderson, 'Engenhos na várzea: uma análise de declínio de um sistema de produçao tradicional na Amazónia' in P. Lena and A.E. de Oliveira (es), *Amazónia: a fronteira agrícola 20 anos depois*, Museu Paraense Emílio Goeldi, Belem, Brazil, 1992; Catherine Matheson, 'Farming the Forest: Sustainable Alternatives in Brazilian Amazonia' in this book, pp)
[10] M.M. Mattos and C. Uhl, 'Economic and Ecological Perspectives on Ranching in the Eastern Amazon', *World Development* 22:2:1994, 145-58.
[11] See S.B. Hecht, 'Environment, Development and Politics: Capital Accumulation and the Livestock Sector in Eastern Amazonia', *World Development* 13, pp 663-84.
[12] Mattos and Uhl 1994, op cit.
[13] Mattos and Uhl op cit, p.18.
[14] Ibid
[15] M.H. Allegretti, 'Extractive Reserves: An Alternative for Reconciling Development and Environmental Conservation in Amazonia' in A. Anderson 1990, op cit.
[16] See Scott Anderson, op cit 1992.
[17] D.G. McGrath, 'The Paraense traders: small-scale, long distance trade in the Brazilian Amazon,' University of Wisconsin-Madison (unpublished PhD dissertation) and 'Fishers and the evolution of resource-management on the Amazon floodplain', *Human Ecology*, 1993 21:2: pp167-95.

Did Chico Mendes die in vain? Brazilian rubber tappers in the 1990s

Anthony Hall

Chronicle of a death foretold

Several years have passed since the fateful day of 22 December 1988 when Francisco Alves Mendes was struck down by an assassin's bullet. As leader of the Xapuri Rural Workers' Union, located in the western Amazonian state of Acre, 'Chico' Mendes had helped spearhead one of the most significant social-environmental movements in Latin America, that of the rubber tappers or *seringueiros*. His murderers, recently arrived cattle ranchers or *fazendeiros* from the south of Brazil who resented the tappers' growing power to resist land-grabbing and the destruction of their rubber stands, were eventually caught, tried and convicted. That the assassins were brought to justice at all was due largely, however, to the massive international campaign which immediately followed the murder. The perpetrators' subsequent escape from gaol came as no real surprise in a country where 99 per cent of homicides arising from rural land conflicts go unpunished.[1]

Chico Mendes had lived in the shadow of death for a decade. At his own fortieth birthday party on 15 December 1988, one week before he was killed, he stoically predicted, 'I don't think I am going to live until Christmas.'[2] Mendes was the ninetieth rural activist to have been murdered in Brazil that year. The previous 89 had produced little or no impact, but his death provoked a massive national and global reaction which has had lasting repercussions for Amazonia. Over 4,000 people accompanied the funeral cortege in Rio Branco on Christmas Day, while the world's news media reverberated with headlines and leading articles examining events leading up to the assassination and highlighting the rubber tappers' struggle to defend their livelihoods.

The intensity of this reaction undoubtedly reflected widespread humanitarian concern as well as solidarity with the cause of the seringueiros. Yet the world's attention had also been focused on Amazonia during the late 1980s as the result of record levels of forest loss and burnings, especially in 1987, barely a year before the assassination. Resulting international protests over the impact of rapid deforestation on global warming and biodiversity loss brought an acrimonious and nationalistic response from the civilian administration of President Sarney, which had been heavily criticized for doing too little to tackle environmental degradation in the Amazon region and the social injustices which invariably formed part of this process.[3] The events of December 1988 were thus 'the last straw', and catapulted the rubber tappers' movement to centre stage in the evolving drama. The Sarney government, rather taken aback by the scale of international condemnation, set up a new environmental control agency (IBAMA) and elaborated Brazil's first national environmental policy, the nationalistic *Nossa Natureza* ('Our Nature'). In addition, the publicity surrounding Chico's death also brought about important changes in public policy for Amazonia, including the setting up of several protected 'extractive reserves' (*Reservas Extractivistas*) for rubber tappers. Ironically, his influence in death was stronger than in life.

The History of the rubber tappers

From the 1870s to around 1912, Brazilian rubber production expanded rapidly in response to growing world demand from, amongst other sources, the nascent automobile industry. Businessmen set up rubber estates (*seringais*) along the major rivers in the states of Amazonas, Rondônia and especially Acre, annexed from Bolivia in 1903, which accounted for some two-thirds of total production at the height of the boom in 1900. Starving peasants fleeing the ravages of north-eastern droughts supplied the bulk of the estate labour force, complementing a small and rapidly dwindling indigenous Amerindian population.

Following the collapse of the rubber boom in the early twentieth century the rubber economy fell into decline, despite government attempts to revive it during World War II, and estates were gradually abandoned by their owners (*seringalistas*). During the 1960s in particular, as the price of Brazilian rubber dropped in the face of growing foreign competition from Malaya,[4] estate bosses switched their attentions to urban-based commercial activities. Traditional ties of debt bondage were cut or loosened, but the old bosses were replaced by itinerant traders (*marreteiros*) who have perpetuated debt dependency through their intermediary role. The average rubber-tapping family has become the effective owner of its own area of rubber stands (*colocação*), now with more freedom to grow subsistence foods and harvest tree crops such as Brazil nuts. Living standards remain low and malnutrition is common due to a starch-heavy diet. Illiteracy is the norm and there is a high incidence of communicable diseases such as malaria and lieshmaniasis.

During the 1970s, the livelihoods of rubber tappers were placed under serious threat by an influx of large landowners and businessmen, mainly from southern Brazil. As part of its policy of 'integrating' and 'modernizing' Amazonia, the Brazilian military regime gave them generous financial incentives. Over US$ five billion was spent on such subsidies for

Amazonia during the 1970s and 80s, mostly for the setting up of cattle ranches; these have been both economically unproductive as well as ecologically disastrous.[5] The pressure of the advancing 'cattle front' induced the rubber tappers in southern Acre (through newly-founded rural trade unions) to actively resist the appropriation of their traditional lands. The *empate* (a stand-off or confrontation) was devised as an essentially peaceful resistance tactic in which men, women and children came face-to-face with labourers hired by logging companies or by ranchers to clear the rainforest. The rubber tappers would simply ask them to stop their activities. The first empates took place in 1976 and by 1988 some 45 had been organized, reputedly saving over one million hectares of forest from destruction.[6]

Although the empate is still used by rubber tappers today, since the 1980s efforts have focused on the design and implementation of a longer-term management strategy. Following the demise of the military regime in 1985 major advances were made in the organization of rubber tappers. With the assistance of a Brazilian NGO (INESC) and Mary Allegretti, an anthropologist who had been associated with the rubber tappers' cause for several years, the first National Meeting of Rubber Tappers was organized at the University of Brasilia in October 1985. There, it was decided to set up the National Council of Rubber Tappers (*Conselho Nacional dos Seringueiros* – CNS) to represent tappers' interests and lobby the new civilian government for policy reform to protect seringueiros' land rights and set up development programmes.

Extractive reserves

The Council also set up a working group to promote the idea of extractive reserves, adapted from the concept of Indian reserves. The CNS hoped that these could be included within the 1985 Brazilian land reform programme. At this time, the Brazilian government had come under intense international pressure over the World Bank-funded North-West regional development programme (POLONOROESTE) and, in particular, the proposed paving of the BR-364 highway to Rio Branco, state capital of Acre. The BR-364 project had already been heavily criticized for stimulating uncontrolled immigration and deforestation in neighbouring Rondônia state. Matters came to a head in 1987, when Amazon burnings reached a peak and attracted international condemnation. A series of violent confrontations took place that year, starting with an empate on the *Seringal Cachoeira*, home of Chico Mendes, against the rancher Darly Alves da Silva.

In order to avoid undue conflict and adverse publicity, the federal government intervened and expropriated four estates to form the first 'Extractive Settlement Projects' under the 1985 land reform legislation. By 1989, ten such projects had been set up covering 900,000 hectares with some 3,000 rubber-tapper families. The murder of Chico Mendes and the resulting global publicity provided the necessary fillip for prompt government action. A few months later, the extractive reserve was legally instituted within Brazil's new National Environment Programme (PNMA) and in 1990 it was formalised by presidential decree. The government was supposed to expropriate designated areas and compensate their owners, after which a demographic census and land-use survey would be undertaken. Through their representative local associations, rubber tappers were to help design and approve an environmental management plan (*Plano de Utilização*) for each reserve, which would serve as the basis for a long-term lease contract between the federal government (through IBAMA) and the reserve population. Rubber stands were to be sub-leased to rubber-tapping families by the local association which would administer the reserve collectively. Land sales to outsiders would not be allowed, while only non-destructive economic activities would be permitted.

Unlike normal Brazilian conservation units, in which no form of productive activity is permitted, the extractive reserve is an attempt to reconcile the conservation of natural resources with their sustainable use in supporting rubber tappers' livelihoods.[7] The first federal extractive reserve was decreed in January 1990 for the Juruá valley in northern Acre, followed in March by three more: the Chico Mendes reserve in southern Acre, Rio Ouro Preto in Rondônia and Rio Cajarí in Amapá. These four units alone cover over two million hectares and are home to a population of around 23,000 people. In addition to these federally-decreed extractive reserves, there has been a movement by the Organization of Rubber Tappers of Rondônia (OSR), separately from the CNS, to press the state government into establishing 28 smaller reserves, covering a further 1.7 million hectares.

Making the extractive reserves viable

Now that the territorial integrity of the reserves is guaranteed by law, a number of other prerequisites still have to be met if the dual aim of conservation and sustaining livelihoods is to be achieved. The successful design and implementation of a long-term management plan will depend on: 1) the generation of higher incomes for rubber tappers – given the declining market price of natural rubber – through a range of diversified economic activities, from production through to marketing, 2) the enhancement of social and political unity amongst rubber tappers to support both production initiatives as well as the efficient governance of common property resources within the reserve, 3) continuing financial, technical and logistical support from the Brazilian government and foreign donors, and 4) policy reform within Brazil to favour the extractive economy.

Of these four prerequisites, funding is the least of their immediate problems. Under the 'G7 Pilot Programme to Conserve the Brazilian Rainforest', over US$ nine million has been earmarked for development of the four largest extractive reserves (Alto Juruá, Chico Mendes, Rio Ouro Preto and Cajarí).[8] These funds are designed to assist the Brazilian Environment Institute (IBAMA) in: (a) carrying out baseline socio-economic and biodiversity studies as well as boundary demarcation, (b) the strengthening of basic infrastructure for production, transport, health and education, and (c) improving the management capacity of the community associations and cooperatives whose ultimate task it is to administer the reserves. In addition, further bilateral and NGO financial support has been given to sub-projects such as the Chico Mendes reserve cooperative (CAEX), discussed below. As a result of their high political profile, extractive reserves in Amazonia have been quite privileged. Yet if funding has been guaranteed at least in the short-to-medium term for start-up operations, other impediments lie ahead on the rocky road to sustainability.

Of the various potential obstacles which could undermine the viability of extractive reserves, perhaps the most immediately pressing is that of economic self-sufficiency. The declining market price of latex has proved a serious problem. Throughout the 1980s, government subsidies for domestic rubber were gradually withdrawn and cheaper Malaysian imports encouraged, while synthetics have become increasingly popular. The market price of latex fell from US$1.80 per kg. in 1980 to under US$0.40 per kg. by 1992. In 1990, a special tax on imported rubber (the TORMB) was withdrawn, reducing funds available for subsidizing Amazonian production, which declined from 85 per cent of national output in 1985 to 28 per cent by 1992. Precarious living conditions on the reserves, together with the lure of other income-earning opportunities such as gold prospecting or within the urban informal sector, are fuelling the gradual depopulation of rubber-tapping areas and the swelling of shanty towns in the region's capital cities.[9] Rubber tappers' incomes need to be diversified and strengthened to provide them with an incentive to remain on the reserves and preserve the forest. Surveys have shown that rubber and Brazil nut gathering still account for two-thirds of tappers' income, complemented by subsistence farming and hunting. Efforts to increase rubber tappers' income have therefore focused on these activities. In 1993, the CNS organized a major demonstration in Brasilia to pressure the government to review the tax issue and create an investment fund for extractive reserves. A government interministerial commission is presently reviewing these matters.

Through federal government training programmes and the provision of appropriate equipment, efforts are also underway to improve the quality of Amazonian latex. Smoked rubber sheets (*Placa Bruta Defumada* – PBD) have fewer impurities and command a price forty to seventy per cent higher than traditional rubber. A new product called 'plant leather' (*couro vegetal*), comprising sheets of cotton coated in rubber which are then smoked and vulcanised, is also being produced and bought by companies specialising in the manufacture of shoes, bags and other consumer items for national and international markets. The cooperative in Xapurí is also planning to build a processing factory to serve 2,600 tappers in Acre, financed through a special credit line from the Bank of Brazil, to produce high-quality rubber for niche markets.

Efforts to increase the value of Brazil nut production in the reserves have proved less successful. In 1989 the Agroextractive Cooperative of Xapurí (CAEX) was set up to serve the Chico Mendes reserve. The US-based NGO, Cultural Survival, provided seed money, while in the space of five years over US$1.6 million was attracted in overseas grants. The cooperative's purpose was to replace exploitative intermediaries in the purchase of Brazil nuts and rubber, thus increasing the profit margin by selling directly to national and foreign buyers, and undermining the monopoly over processed nuts held for generations by the Mutran family.[10] In addition, CAEX was to set up a consumer cooperative for basic items.

However, major problems soon became apparent. Production at the Xapuri plant was undermined by high rates of absenteeism and the use of inefficient technology for opening and processing nuts. Severe overstaffing at the factory and a heavy social security bill inflated costs and left CAEX exposed to competition from Bolivia, currently the world's largest producer, where the price of nuts was forty per cent lower. To address these problems, nut production was decentralised to four rural mini-factories (the *Projeto Castanha*). Ostensibly it was claimed that these mini-factories would reduce transportation costs and wastage, but the prime reason seems to have been to cut the social security bill by switching from a waged labour force to a self-employed, piecework system. Decentralisation has not proved very successful. The mainstay of the nut-processing labour force were women who were unable to dedicate themselves to the mini-factories during the early phase of the processing cycle due to competing demands on their time from agriculture. Although children were mobilised as a substitute, their productivity was lower and their participation prevented them from attending school. In mid-1994, CAEX ceased Brazil nut processing and, ironically, sold the harvest to its arch rivals, the Mutran company. Mismanagement and poor planning had not only seriously undermined the CAEX enterprise but had also knocked the confidence of the rubber tappers on the Chico Mendes reserve.

Whatever potential there is for boosting incomes through rubber and Brazil nut production, economic success may also depend on diversification into less traditional activities such as agro-forestry and other non-timber forest products (NTFPs), and perhaps selective logging. Here, too, there are potential problems as the market potential for alternative NTFPs is largely untested either in Brazil or overseas. Furthermore, the CAEX experience illustrates the possible pitfalls in administration and processing (see Stephen Nugent and Catherine Matheson in this book).[11]

Apart from economic sustainability, the success of extractive reserves will also hinge on the capacity of rubber tappers to collectively manage their territory. Above all, rubber tappers need to feel confident that forest preservation serves their long-term interests. In the past, threats from loggers and land-grabbers made rubber tappers feel insecure about their tenure and induced some to engage in slash-and-burn farming or abandon their land to cattle ranchers for short-term gain. It is hoped that legal recognition of land rights, as well as a system of permanent vigilance of reserves to monitor illegal incursions, will discourage these practices and avoid a 'tragedy of the commons' scenario. Under the 1990 law, collective USufruct rights (*Concessão Real de Uso*) were granted by the federal government to rubber tappers' associations established to run each reserve. These associations, in turn, can grant formal tenure to individual families, allowing them to carry out extractive activities but not permitting resale of the land to outsiders. By late 1995, six reserve associations had been set up, each with revolving funds for the purchase of Brazil nuts, rubber and other NTFPs from reserve inhabitants. In collaboration with IBAMA, reserve protection and surveillance plans have also been developed.[12]

It can be argued that no group is better placed to administer the reserves than the tappers themselves and their representative associations, given their presence on the ground and vested interests in seeing the idea work. However the potential problems should not be under-estimated. Firstly, the sheer size of these territorial units coupled with a low population density and lack of transport or roads makes communication extremely difficult. The Chico Mendes reserve, for example, is almost one million hectares in size, and has barely 12,000 inhabitants. But fragmentation is as much socio-political as it is geographical. Relations amongst rubber tappers have been characterized by vertical ties with powerful patrons (in earlier times the rubber estate owners or *seringalistas* and more recently intermediary merchants or *marreteiros*). Despite the recent history of mobilisation against land-grabbing through empates in some areas such as southern Acre, rubber tappers lack a tradition of collective or class-based action. Overcoming this history of physical isolation and individualism in order to manage the reserves through collective associations is a major challenge in the quest for sustainability.

In designing and executing management strategies for extractive reserves, their diverse social structures and histories have to be taken into account. The Chico Mendes reserve in southern Acre, for example, despite its huge size, enjoys relatively high levels of participation in its assocation due to its recent history of collective action against invaders. The Juruá reserve in northern Acre, on the other hand, has remained far more isolated and is not yet connected by tarmac (only by a dirt road) to the national highway network. Social relations have been more heavily conditioned by traditional debt bondage and the area has no history of confrontation, making it more difficult for the newly formed association to recruit members.[13]

Aside from the organizational effectiveness of each extractive reserve is the problem of maintaining the political unity of the rubber tappers' movement regionally and nationally. The Rubber Tappers' Council (CNS) was born of the early struggles against outside land-grabbers. Now that the movement has matured and has entered the political arena, the degree to which it genuinely represents rubber tappers' interests has been called into question and its initial unity of purpose has perhaps become undermined. The CNS now claims to represent all extractivists and even landless groups across Amazonia, not just rubber tappers. This switch reflects pressure from NGOs, the Brazilian Workers' Party (PT) and the associated trades union federation (CUT). It has created tensions between traditional rubber-tapping groups in western states and newer activists from eastern Amazonia who support the landless and agrarian reform on a wider scale. Institutional rivalries and personal clashes within the tappers' 'movement' have further exacerbated internal divisions since the early 1990s.

Finally, the viability of extractive reserves in the long term will depend on the creation of a favourable policy environment in Brazil for extractivism. Under military regime and civilian rule alike, development subsidies for Amazonia in the form of tax breaks and special credit lines have favoured large-scale, commercial enterprises in livestock, timber and mining. Extractive and other traditional technologies have been almost totally ignored. The government of Fernando Henrique Cardoso has recently outlined a new development strategy for the Amazon region which, for the first time, explicitly recognises the legitimacy of extractive and other indigenous activities in terms of both their economic and their broader environmental value.[14] It remains to be seen, however, whether this will be translated into long-term political and financial support for reserves.

Conclusion

The struggle of the rubber tappers and their allies to create extractive reserves represents the single most important policy innovation of the 1980s and 1990s in favour of Amazonia's traditional inhabitants. Yet it would be foolish to romanticise the seringueiros' movement which has yet to demonstrate either the economic viability of reserves or their feasibility as units which can be managed successfully by local populations. At the end of the day, seringueiros are no more inherently environmentally friendly than any other social group which depends on exploiting natural resources for its survival. The incentive and management structures must be created for conservation and production to become realistically compatible and solutions must be adapted to fit the diverse range of situations encountered in sub-regions of Amazonia. The success of extractive reserves will depend on whether the tappers' organizations, together with development practitioners, policy makers and other participants in this process, can create the necessary conditions. Only then, perhaps, can we say that the death of Chico Mendes was not in vain.

[1] Darly Alves da Silva and his son Darci were sentenced to nineteen years in prison but escaped in mysterious circumstances shortly after the trial. They are currently reputed to be hiding across the border in Bolivia.
[2] A. Revkin, *The Burning Season: The Murder of Chico Mendes and the Fight for the Amazon Rain Forest*, Collins, London, 1990, p.273.
[3] For further details on environmental policy debates for Amazonia during this period, see 'Introduction' in D.Goodman and A.Hall (eds.), *The Future of Amazonia: Destruction or Sustainable Development?* Macmillan, London, 1990.
[4] The story of Britain's role in the illegal transfer of rubber seedlings to Kew Garden (with the aid of a British landowner based near Santarém, Henry Wickham) is well known. These were transplanted to the British colony of Malaya, whose successful rubber plantations undermined Brazil's supremacy in this field. For an excellent account see: Warren Dean, *Brazil and the Struggle for Rubber*, Cambridge University Press, Cambridge, 1987.
[5] The regional development agency SUDAM had approved 631 livestock projects in Amazonia by 1985. Only 15 per cent of these schemes are officially considered successful, the remainder being fronts for land speculation and related, non-approved activities. See D.Mahar, *Government Policies and Deforestation in the Amazon Region*, World Bank, Washington D.C., 1989.
[6] Chico Mendes with A. Gross, *Fight for the Forest: Chico Mendes in His Own Words*, Latin America Bureau, London, 1989.
[7] For a critical appraisal of ICDPs, see C.Barrett and P.Arcese, 'Are Integrated Conservation-Development Projects (ICDPs) Sustainable?' *World Development*, 23 (7), July 1995, pp. 1073-84.
[8] The G7 Pilot Programme is a US$250 million initiative which was announced at the Houston meeting of the G7 in July 1990. It channels grants from a central Rainforest Trust Fund as well as bilaterally pledged funds and the Brazilian government to several sub-components, including extractive reserves. See: Friends of the Earth, *Sound Public Policies for the Amazon Region*, São Paulo, 1994 and A.Hall, *Making People Matter: Development and the Environment in Brazilian Amazonia*, Occasional Papers No.4, Institute of Latin American Studies, University of London, 1993.
[9] In Acre official figures suggest the extractive population declined by 17 per cent from 1960-80.
[10] The Mutran family's business empire is based in Belém and Marabá, in Pará state.
[11] For cautious optimism about the future of extractive reserves see: A.Anderson, 'Land-use Strategies for Successful Extractive Economies' in Friends of the Earth (ed.), *The Rainforest Harvest*, London, 1992, pp. 213-223, and S.Schwartzman, 'Mercados para Produtos Extractivistas na Amazonia Brasileira', in R.Arnt (ed.), *O Destino da Floresta: Reservas Extrativistas e Desenvolvimento Sustentável na Amazônia*, Relume Dumará, Rio de Janeiro, 1994, pp. 247-257. For a more sceptical analysis, see: J.Browder, 'Extractive Reserves and the Future of the Amazon's Rainforest: Some Cautionary Observations' in Friends of the Earth, op cit., pp. 224-235, and A.Homma, *Extrativismo Vegetal na Amazônia: limites e oportunidades*, EMBRAPA-SPI, Brasilia, 1994.
[12] *Rain Forest Pilot Program Update*, World Bank, Washington D.C., 3 (4), October 1995.
[13] Observations based on author's fieldwork during 1994-95.
[14] *Politica Nacional Integrada para a Amazônia Legal*, Ministerio do Meio Ambiente, dos Recursos hidricos e da Amazônia Legal, Brasilia, July 1995.

Fruit farming in the Brazilian Amazon: a sustainable alternative

Catherine Matheson

The world watched in horror as the Amazon burned in the 1980s, shocked that future generations could be deprived of an irreplaceable international legacy. But for the Brazilian farmers who live in the Amazon the problem is one of immediate survival: how to make a living without destroying their environment. Amazonian inhabitants suspect Western motives, as one agronomist explains: 'The Europeans have a romantic idea of the Amazon. They see the Indians wearing their loin-cloths and they think we are in Paradise. They need to get out of their offices and come and see how difficult it is to live in the forest. It's a hard life...'[1]

In eastern Amazonia a group of active peasant unions involving thousands of farmers has joined with university academics and technicians in an innovative initiative to live in harmony with the rainforest. The project, which is funded by Christian Aid, the European Union and the British government, among others, is based at the Agricultural and Environmental Centre of Tocantins (CAT) near the town of Marabá in Pará state. It grew from the farmers' own recognition of their dilemma: that their habitual cultivation methods were destroying their livelihood, as well as that precious legacy of future generations, the Amazon rainforest. They are a group of particularly aware and active farmers with a history of collective struggle arising from their bloody battle for land rights. But their trajectory has a number of useful lessons for other colonizing farmers.

Across the Amazon as a whole, most environmentalists agree that the principal cause of deforestation is large-scale cattle ranching.[2] In many cases, however, peasant farmers cultivated annual crops before grass was planted. Their expansion of the agricultural frontier prior to cattle ranching has been one of the major factors in deforestation.[3] One estimate says there are about five million people dependent on semi-subsistence farming in eastern Amazonia alone[4] and each farmer burns on average one hectare of forest per annum to cultivate crops. Like colonizing peasants everywhere, Brazilian migrant farmers use slash-and-burn to clear rainforest for farming. However, without its tree cover, rainforest soil quickly loses its fertility, forcing farmers to convert to pasture for cattle and then move on when even fodder will not grow. For technicians and environmentalists, the issue may be to preserve the Amazon or 'to stabilise the colonization zone', as one technician said, but for the farmers themselves, the issue is self-preservation for them and their families. Raimundo Vieira da Silva, who has worked the land in eastern Amazonia since he was thirteen years old, explains the problem: 'We have to burn the forest down because if we didn't, we would need a machine to do the job before we could make this land produce. To survive, you have to burn down some forest every year.'[5]

In response, the project which farmers, academics and technicians have formulated together has the double objective of limiting the burning of the Amazon and allowing farmers to stay on their land. The project has benefited from the formation of the Farming Foundation of Tocantins-Araguaia (FATA) which united the five local branches of the rural workers' union. This has provided the project with a pioneering bedrock of 12-15,000 families who are now experimenting with the cultivation of permanent crops such as fruit and nut trees. To work with the farmers, academics specialising in agronomy, sociology, economics and anthropology have formed a research group called the Socio-Agronomic Laboratory of Tocantins (LASAT). This innovative collaboration has not been without its problems, but has largely proved a productive way forward.

FATA has its origins in the migration of hundreds of thousands of poor migrants from other parts of Brazil into eastern Amazonia in the 1960s and 1970s, as the building of roads, railways and dams breached one of the natural world's last frontiers. They cleared the rainforest by burning it and then had to confront the hired gunmen of large landlords and land speculators who were also after the Amazon's riches. In the state of Pará 103 people were killed in land conflicts in 1985 alone.[6] The area around the town of Marabá saw much violence, situated as it is near the centre of the Grande Carajás Progamme, the giant Brazilian government industrialization programme for the eastern Amazon. In the twenty years up to 1988 the town's population rose from 10,000 to 200,000; the region became the Amazon's agricultural frontier with the construction of the Belém-Brasília highway and the Transamazonica highway. For the Brazilian government, the Amazon was a 'safety valve' for diffusing social pressure caused by landlessness in the north-east and the south.[7]

In June 1985 a notorious massacre took place on a former Brazil nut estate called Uba, where thirty families had occupied part of the 4,000-hectare area owned by Edmundo Virgulino. Seven people, including a woman who was three months pregnant, were killed by gunmen as they worked the land. Five days later the gunmen returned to search out the community leader, José Pretinho. His widow who was eight months pregnant at the time, tried in vain to warn him but he was shot five times and died, leaving her with five children.[8] The leaders and survivors of this bloody struggle founded FATA and their shared suffering has given them the unity on which to build their crop experimentation and marketing activities. Raimundo Vieira explains: 'Those friends of ours who died, we see them as seeds that didn't die but which are still living. We had to make their dreams come true.'

FATA leader Francisco de Assis joined the union at fifteen. At this tender age, he was leading a group of 28 farmers when they were threatened by the same gunmen who killed Pretinho. For Francisco, the founding of FATA is a continuation of the collective struggle for land rights which has now moved on to ensuring the security of livelihoods for future generations. 'It's a question of organizing people to work together to achieve common goals,' he says.[9] FATA members like Francisco are driven by a quasi-spiritual, religious inspiration which has historical roots: the churches provided help and support for community groups in the Amazon and elsewhere in Brazil at a time when other progressive organizations had been driven underground by the military government. Today, several of the leadership positions in FATA and CAT are filled by priests and lay preachers.

FATA coordinates the village groups which are setting up tree nurseries for experimental species. These groups are supported by agronomists and technicians who travel around on motor-bikes, by boat or by foot. Regular meetings are held at the Agricultural and Environmental Centre between representatives of the scientists, technicians and farmers to assess progress and discuss plans. The Centre also operates a nursery for different tree species, especially fruit and timber.

The species being grown include exotic fruits such as passion fruit, papaya, mangosteen and others without English names including cupuaçu (*Theobroma grandiflorum*), acerola (*dicotyledon*), araçazeiro (*Psidium littorale*), and biribá (*Annona lanceolata*). In addition, experiments are underway to combine traditional production techniques with modern ones. For example, the genipapo tree (*Genipapa americana*) is a shade-giving tree whose fruit is used for making drinks. Amazonian Indians use its juices, mixed with charcoal, for body painting. It is being grown on steep slopes with pigeon pea (*Cajan cajan*), a leguminous ground cover traditionally grown in Asia which provides animal fodder and prevents soil erosion. The cupuaçu fruit, which has yellow pulpy flesh and is used for ice cream and desserts in Brazil, has been the most successful. It was the farmers' idea to grow the shade-loving cupuaçu in combination with Brazil nut trees because traditionally they had been grown together. A cupuaçu produces 35-50 pieces of fruit per annum after four years, whereas a Brazil nut tree takes twenty years to produce nuts.

The issue at the heart of this sustainable agricultural model is finding a market for the produce. In contrast to other Amazonian products such as Brazil's mahogany, of which the UK is one of the largest importers, the fruits of the forest are not in great demand overseas. Consequently, the project must orientate itself towards the local market. To these ends, FATA has set up a cooperative for processing, storing and selling cupuaçu locally (as well as Brazil nuts, to a lesser extent). The local market now buys all the cupuaçu the farmers can supply. Recently the cooperative has also started to buy cupuaçu from the Gaviaô indigenous community who live in the nearby Mâe Maria reserve. Although there has been some conflict over land between migrant farmers and Indians, it is worth noting that the Gaviôes too have a history of resistance to powerful interests, perhaps stemming from their strong warrior ethos. In 1987 they twice threatened to block the railway taking iron ore to the coast if settlers were not evicted from their land.

Now boasting a membership of 770 members, the FATA cooperative does not only market cupuaçu and Brazil nuts; it also organizes trade in more traditional products such as rice, beans, and flour, in order to maximise the profits to local farmers. As cooperative president, Almir Ferreira Barros, explains: 'The big problem here is that middlemen buy up produce at a very cheap price and there's not very much in it for the farmer. We lend them money at harvest-time, then transport and store the produce until we can get a good price for it.' However, in future, cooperative members are intending to cut down on rice sales because it takes too long to sell at a decent price. Renilde Santana da Silva, a single parent with three children, for example, grows rice alongside her beans, bananas and fruits. In 1994 she sold ten of her 160 sacks of rice to the cooperative at a good price but in her view the cooperative should be able to handle more.

Looking to the future and the importance of educating children about the environment, an education team based at the Agricultural and Environmental Centre of Tocantins runs courses for local teachers in how to teach environmental awareness. A teacher in the village of Sapecada, Dolores Costa, who is herself growing cupuaçu and other fruits, has plastered the walls of her classroom with drawings of local trees, flowers and animals. 'If we don't teach the children how to preserve the forest we still have, then in a few years we won't have any left at all. We will suffer the same fate as the Brazil nut trees – and there's none of those left here.'

The lessons of FATA lie in the methods of its design and implementation, invalidating the stock argument of development policy-makers that small-scale projects are insignificant because they are not replicable. FATA's methods could be replicated elsewhere if they were adapted to local circumstances. Its vision offers a real alternative to the orthodox model of development because: it was conceived with the participants' involvement; it had family needs as the starting point; the farmers themselves are implementers; a local market is actively being sought and sustainability is a vital consideration in the project. Unlike many orthodox projects, FATA addresses the wider issues which are essential for sustainability such as marketing, sources of credit and environmental education.

FATA has a holistic approach which brings together technicians, farmers and academics as partners in the projects. This combination of testing academic theories with the people who will be most affected by them is also being implemented in western Amazonia. The activities and research being carried out at FATA suggest 'there is much more scope for the widespread practice of agro-forestry than had previously been thought,' according to one study.[10]

Agronomist Aluízio Solyno, who works at CAT, insists: 'We don't think that slash-and-burn should be treated as taboo. It is the main tool of the peasant economy. The important thing is to find a way of using fire and forest in a way which takes account of the farmers' needs and the demands of the scientists with regard to the diversification of species. We have to look at ecology from a social standpoint. The most important thing in the forest is the people who live there.'[11]

[1] Author's interview with Aluízio Solyno, agronomist at the Agricultural and Environmental Centre of Tocantins (CAT), June 1994

[2] Anthony Hall, *Developing Amazonia: Deforestation and Social Conflict in Brazil's Carajás Programme*, p.145, Manchester University Press, 1989; Dennis J. Mahar, *Government Policies and Deforestation in Brazil's Amazon Region*, p.8. IDRD/World Bank 1989.

[3] Dennis J. Mahar, op.cit., p.7

[4] Anthony Hall, op.cit.p.259.

[5] Author's interview, June 1994.

[6] Alfredo Wagner Berno Almeida, *Carajás: a guerra dos mapas*, p.286-292 Falangola, Belém, 1994.

[7] Hall, op cit, 1989, p.152.

[8] Author's interview with victim's widow, Marina Ferreira da Silva, May 1994.

[9] Author's interview, June 1994.

[10] Hall, 1989, op.cit., p.261.

[11] Author's interview, June 1994

Pioneer women and the destruction of the rainforests

Janet Gabriel Townsend

Women as green guerrillas?

Across Latin America, the increasing concentration of land in the hands of a few has driven millions of peasant families towards forest frontiers. This migration of 'pioneers' (also described as 'colonists' or 'settlers') has contributed to the heavy toll exacted on Latin America's rainforests in recent decades. The destruction has been underpinned by the traditional slash-and-burn techniques pioneers use to clear forest for agriculture.

The question I shall explore is whether women pioneers have any answer to the destruction of the rainforests. Ecofeminists such as Vandana Shiva (1989) believe that women are closer to nature than men and that they are more aware of the needs of nature, of the needs of their children and grandchildren and of the links between these needs. Can women pioneers be drawn into the ranks of green guerrillas to protect the future? In Brazilian Amazonia, some women have certainly resisted the charcoal burning along the Carajás Iron Ore Railway and others are active in indigenous green movements and in the crusade for extractive reserves. But what of the majority? Women are a diverse group: do they have anything relevant in common?

'If we don't cut down the trees, what will there be to eat?' asks Cristina, a pioneer woman in southeast Mexico.[1] Cristina, alas, speaks for most Latin American pioneer women living in the rainforests whose common goal is a livelihood and who are pioneering for want of a decent alternative.[2] Thirty years after her first hard days in the wilderness, Cristina is lucky. Her husband still has the farm they created and she now lives in a town, a mere hour's drive from a paved road; she runs a small roadside eatery and sells fruit. 'My children will not suffer as I suffered.' Few pioneer women can achieve this, as pioneer families are frequently driven off their land by force or by failure, or secure only a meagre livelihood.

Like many other women pioneers, Cristina has never had intimate contact with the land. Out of 75 pioneer women studied in the Serranía de San Lucas, Colombia, only three worked on the land.[3] There are places in the Mato Grosso, in eastern Colombia and in southeastern Bolivia, where women do work in the fields and forest, but more often pioneer women become forest-dwelling housewives, working long days to provide children, sexual services, food, laundry, hygiene, childcare, health care, garden and farmyard produce and often water and fuel (all unpaid). Many women pioneers have told me of their problems, but none ranked the destruction of the forest as high as poverty or the lack of education and health care. What does concern them, as we shall see, is what will happen when the forest runs out and no more food can be grown.

Pioneer women's problems

Worldwide, women who become pioneers commonly lose rights and income, work harder, have poorer access to services and lose their social networks.[4] A study conducted by five Mexican women and myself in southeast Mexico (in which we interviewed two hundred pioneer women) confirmed this picture. We heard little about deforestation. Most women spoke of increasing poverty, of their lack of public services and of domestic conflict and violence, all of which are caused in part by pioneering, as it combines migration with environmental change.[5] Because these are new farms and new settlements, new public services have to be screwed out of the government or paid for privately, whether they be electricity, piped water, clean water, schools, health care or a road to market. They may take decades to obtain and their absence makes domestic work and childcare very hard. On top of this, rainforest pioneers have to cope with new diseases and with heat and damp which may be unfamiliar to them. The change in diet can also cause severe malnutrition in their children.[6]

Women pioneers are invariably isolated. Having come from hundreds of miles away, they have usually lost all their friends and relatives and are socially naked. Men pioneers must get to know people for the sake of the farm, but it may not be respectable for a woman to do so. In Colombia, I interviewed women who had delivered their babies alone (even though the custom was to have a mother or midwife present) 'because I didn't know anyone to ask'. Often, Colombian women pioneers do not know the woman on the next farm. Both drink and violence are common problems in rural Latin America, but, with agrarian crisis, the economic pressures on men are greatly increased, conflicts around birth control are likely, and women tend to lack support. Certainly in southeast Mexico women see domestic violence and marital rape as leading problems, which we had not expected, or had not expected them to express.

These villages were facing the agrarian crisis all too common to people trying to settle tropical forest in Latin America: the change from food crops to cattle, also known as 'the grass revolution of the American tropics' or as 'hamburger and frankfurter imperialism'. Pioneers trying to make farms out of the forest usually cut it down, burn the debris and plant crops in the ashes. When the fertility is lost, or the weeds take over, or both, they can rarely afford to buy chemical fertilizers and herbicides, so once the forest has been used up they must move on or change from crops to cattle. While few pioneers can afford cattle, they do lay down pasture and sell to ranchers, or rent their land out, or become labourers. Felling the forest and growing crops (without bulldozers or tractors) both require an immense amount of labour. When cattle farms or ranching take over, the demand for labour collapses, and 'the cattle eat up the people'. In Chiapas, Mexico, to cultivate one hectare of maize you need 35 working days, for one hectare of cocoa 80 days – and for one hectare of pasture with cattle, 3.5 days a year.

When pioneers arrive in the forest, workers are at a premium and so are children: 'The boys are our labourers.' During this period, woman's role is thus to bear and bring up a large family: we met many women in their thirties whose husbands would not permit any form of contraception. To become proficient in her role as a mother and housewife, the pioneer woman must learn traditional skills.

But all this changes with the onslaught of cattle ranching which places new demands on women and brings painful conflict in its wake. Above all, cattle require investment rather than labour. This may explain why some of the young women we met had been sterilised after two children, at the behest of their husbands. Alicia, an illiterate girl of sixteen, for example, had had only one child but was to be sterilised after the next. Supporting our findings, Lourdes Arizpe[7] has found that the average number of children in pioneer families in south east Mexico depends on the amount of forest left (as I found in Colombia); the more forest, the more children.

Cattle ranching can bring about a revolution in family economies. Cattle must be sold, food must be bought, as little can now be grown. Many women regret the change, for when the family ate its own maize, they simply fed everyone out of the maize store. Now that the men generate income by selling cattle or by earning a wage on a cattle ranch, women have to extract the money from the men in order to buy food. Poverty again becomes a day-to-day reality, as it was in the first days in the forest. Thus, the switch to cattle may put pressure on women to acquire new skills and/or formal education. This is a problem as a large proportion are uneducated, often having left school to care for the parental family, 'because my mother was ill.' Although some prosperous women have become successful ranchers,[8] pioneer women can rarely get paid work with cattle themselves. The zebu cattle which dominate the American tropics are believed to be savage and therefore 'men's work'.

Agrarian change, then, brings a loss of jobs, often a loss of income, and a change in family size. In a sense, the problems for Latin American pioneer women are less at the first destruction of the forest than at its exhaustion and the arrival of cattle which follows.

Pioneer women's attitudes towards the rainforest

In spite of the traditional view in Spanish America that forest is the opposite of civilization (this view is less noticeable in Brazil, in my opinion), most Mexican pioneer women we met were alarmed about the loss of the forest and only one in ten was pleased to see the forest destroyed (either because it 'shelters insects, snakes and dangerous animals' or because it 'brings thunderbolts'). Some women would willingly leave the forest alone and leave for the city tomorrow, if only they could be sure of a future there. Most want to see alternatives, either better livelihoods from land already cleared, or more off-farm livelihoods for their locality. Mexican pioneer women speak with longing of homeworking and fruit packing for multinational companies – work which is seen as so exploitative elsewhere in Mexico.

Those who were concerned about the depletion of the forest feared that:

- there would not be enough rain (sixteen)
- there would be no land to cultivate (fifteen)
- the animals would be lost (a source of food – twelve)
- there would be a lack of oxygen or clean air (ten)
- there would not be enough firewood (five)
- there would be no wood to build houses (five) and
- the beauty would be lost (five).

They also feared the loss of knowledge, shade and fine timber, and possible increases in disease and drunkenness. We suspect that the worries about oxygen may stem from visits to cities and from TV programmes about pollution in Mexico City, which is certainly appalling but far from the rainforests! In the most remote villages which could only be reached after a long walk, women saw land as central to their children's future and were more concerned than others about whether there would be land to work once the forest was gone. But water topped the list of fears. Across southeast Mexico, pioneers are convinced there is less rain than when they arrived, even though meteorological data show no change. People believe the trees bring water. María de Jesús, aged 85, told us,

> Now we're all convinced by what they say, that the day we felled the trees will be a shame to us. Many have even wept with this weather because it hasn't rained. And it was all because they cut down many trees. The forest attracts the water, so it was their fault.

It is possible that some pioneer women told us they were worried about the loss of the forest because they thought they *ought* to say this to us. It is interesting that only a few believed it was up to women to protect the rainforest because they are more aware. Only two out of forty-five were explicitly ecofeminist and saw women as potential saviours (both are illiterate):

> Women have to watch out that the men don't use up all the forest (Carmela).

> As a woman, you think about the welfare of your children, and if they cut down all the trees, they'll leave all the forest destroyed and a desert. The women want to see a better future for all of us, so that we may breathe clean air (Catalina).

But while Carmela and Catalina believe it is feminine to show sympathy with the forest and its animals, Isabel clearly does not:

> We saw jaguars, big rodents, armadillos, raccoons, badgers – then wild boars and, in the streams, shrimps. We were scared because we didn't know about that kind of jaguar, or the pumas or pheasants... But after a while we took control of the forest and went hunting by night or by day. I had an eighteen-shot rifle – I loved shooting at animals, killing pigeons, doves. Once I killed a small jaguar of about forty kilos. It was in a tree. The chickens were panicking and I went [to see what was going on]... We had to be careful underfoot because of the snakes. There were lots of bushmasters.[9] I don't know how many of those I killed. I killed them with the rifle or with a machete. Just little ones but they still count, don't they?

Clearly, the views of women pioneers are as varied as those of any other women.

Insiders' solutions

Indigenous people, long-term forest residents and outsiders have all proposed alternative uses of the forest to yield better livelihoods, as discussed elsewhere in this book.[10] In Mexico, we were fascinated by the possibilities presented by forest gardens which some of the pioneers had created. This is the one sustainable form of production practiced on the lowlands of southeast Mexico. In the gardens of a single village in Los Tuxtlas, Mexican botanists have recorded 338 different species of plants, of which all but 18 were being used for a specific purpose.[11] These gardens imitate the rainforest in structure, diversity and often appearance: it looks like complete chaos, but everything is carefully judged. The imitation rainforest needs very low inputs to yield steady production. It protects the soil against erosion and requires little or no fertilizer or pesticide and relatively little labour. The pioneers had brought plants with them from elsewhere but had also incorporated plants from the forest to meet a huge range of subsistence needs: food, forage for animals, medicines, herbal teas, glues, timber and ornament. These gardens were a family affair, where men were responsible for the trees, women for the lower-growing plants and the chickens and pigs, children for finding new species to test. But all was under the woman's management.

Now that cattle, milk and the cash economy dominate production, many gardens are shrinking and are neglected. Today camomile tea is often bought rather than picked. Given the interest in organic farming and forest-friendly production among so many consumers in North America and Western Europe, this seems to us a tragedy. In most small towns in the American tropics, cafés sell superb fresh fruit drinks mixed with water or milk: fruits well-known outside the region such as mango or citrus, and the less familiar curuba, tamarind, pitihaya. If these fruits ever reach us here in the North, they are from chemical-hungry, energy-intensive plantations. By contrast, women pioneers are producing organic, low-input, forest-friendly tropical fruits, as well as teas and other commodities. Such work would be attractive to them, for decent Mexican rural women are expected to be in their home and garden, caring for their children, and maintaining their husband's status. There are strong traditions of contract production and cooperative marketing and the gardens could be scaled up to very substantial production. But at the moment, the women know they cannot sell enough to make it worth expanding their gardens. Moreover investment would be needed in canning or freezing factories capable of handling a variety of produce over the year. Understandably, the women see such investment as a Utopian dream. The problem is that by the time there is an adequate international market for the gardens' produce, the knowledge and skills may have been lost.

Worldwide, outsiders' efforts to impose conservation have tended to fail unless local society itself values that which is to be conserved. If Canada cannot protect the cod on the Grand Banks nor the United Kingdom the herring in the North Sea, it ill becomes us to encourage anyone else to impose conservation. If we accept that the solutions need to come from insiders, then the skills and experiences of women pioneers could play a vital role in the search for those solutions.

[1] Cristina, a pioneer woman in Cuauhtémoc, Mexico, interviewed in 1990.
[2] Cristina does not speak for indigenous women or for Amazonian *caboclas* whose families have lived in the forests for generations and who may have different goals.
[3] J.G. Townsend and S. Wilson de Acosta, 'Gender roles in the colonization of rainforest: a Colombian case study' in J.H. Momsen and J.G. Townsend (eds) *Geography of Gender in the Third World*, Hutchinson, London, 1987.
[4] J.G. Townsend with U. Arrevillaga, J. Bain, S. Cancino, S.F. Frenk, S. Pacheco and E. Pérez, *Women's Voices from the Rainforest*, Routledge, London, 1995.
[5] Townsend et al, 1995, op cit.
[6] Townsend and Wilson de Acosta, 1987, op cit
[7] L. Arizpe, F. Paz and M. Velásquez, *Cultura y cambio global:percepciones sociales sobre la desforestación en la selva lacandona*, Miguel Angel Porrua, Mexico City, 1993.
[8] Oddly enough, in eastern Colombia, women work with zebu cattle as successfuly as with dairy cattle in the Andes.
[9] Large, aggressive, poisonous snakes.
[10] See Stephen Nugent, 'Amazonian Indians and peasants: coping in the age of development', Anthony Hall, 'Did Chico Mendes die in vain? Brazilian rubber tappers in the 1990s', and Catherine Matheson, 'Fruit farming in the Brazilian Amazont: a sustainable alternative' in this book.
[11] M.E. Alvarez-Buylla Roces, E. Lazos Chavero and J.R. García-Barrios, 'Homegardens of a tropical region in southeast Mexico: an example of an agro-forestry cropping system in a recently established community', *Agroforestry Systems* 1989 8: pp133-56.

Can ecotourism save Ecuador's threatened cloud forests?

James Fair

From the Bellavista cloud forest reserve in the Pichincha province of north west Ecuador it is a three hour walk to the nearest village of Nanegalito. As you leave the reserve, walking along a rough road which was once a main thoroughfare but now sees no more than one or two vehicles a day, you are as likely to see the endangered plate-billed mountain toucan as you are any other humans. Gradually, however, you begin to see evidence that there are indeed people living in the region. After one kilometre, Finca Ensueño hoves into view, and near the top of the hill, you start to see the first evidence of serious deforestation. Soon, the forest has given way almost entirely to cattle pasture, and in some places there are bare patches where neither trees nor grass grow. Here, the harsh equatorial sun has turned the earth into a caked, red mud – a visible warning of what can happen to the fragile Andean soil when it is stripped of the organic matter that once gave it life. Not only does the soil become infertile and useless, but it can also cause landslides, making roads impassable.

That this is not a huge problem in the hills around Bellavista may be because – near the top at least – they retain much of their forest cover which binds the soil together. The Bellavista reserve – which has no official status with Ecuador's government – is the project of Richard Parsons, a British teacher and ecologist who has lived in the country since 1988. His idea is to preserve the forest by exploiting it through the outlet of ecotourism. He would like to see local peasants benefit from the forest, rather than have to cut it down to grow crops, or more probably in this region, to turn it into pasture. So far, however, he has had little success in involving local communities. To most of them, Parsons admits, he is just a 'crazy gringo', and although two Ecuadoreans are now working for him in the reserve, they are not from the immediate region. Nevertheless, Parsons believes that ecotourism could provide an alternative, or supplementary income to local peasants in a way which is more sustainable than resorting to the chain-saw.

One local farmer who has benefited from Parsons' project is Manuel Ipo, partly through odd employment opportunities, but also because at the end of 1995 he agreed to sell his land to Parsons. Ipo has bought another plot of land lower down in the Andes where the climate is, he says, more suitable for the cultivation of cash fruits like bananas and pineapples. Ipo is enthusiastic about the idea of ecotourism but has little faith that local people will ever change their attitudes or ideas. 'They only want to exploit the mountains for the trees,' he says.

Does it have to be like that? Is the conservation of Ecuador's cloud forests necessarily antagonistic to the economic well-being of the communities that live in and around them. The few remaining patches of forest in these hills are poorly protected by national laws and their plight has gone largely unnoticed in both the United States and Europe as well as in Ecuador itself. Mass colonization of previously virgin areas of forest in both the highlands and lowlands began with the Agrarian Reform Law of 1964. While the original idea was to redistribute land to peasants tied to the feudal *huasipungo* system, it actually resulted in their being encouraged to colonize previously uncultivated areas. José Vicente Zevallos notes that between 1964 and 1985, over 2.5 million hectares of colonization areas were granted to about 60,000 families. While the vast proportion of the colonized land was in the Amazon, during the 1960s, 45 per cent was in the highlands.[1] With probably less than ten per cent of Ecuador's cloud forests still standing, they are arguably more at risk than the rainforests of the Amazonian basin.

Nevertheless, should not people take precedent over plants and animals? If they need to cut the trees, it seems rich for outsiders, gringos in particular, to tell them it is wrong. But any conservationist will say that constant deforestation of the Andes is both wasteful and unsustainable, and ultimately not in the best interests of the people themselves – or certainly not in their children's interests. In 1992 Conservation International warned: 'Destruction of the remaining forests of south west Ecuador will – without any doubt – have a strongly adverse effect on the regional economy. This potentially huge problem should be viewed as a threat to the national security of the country.'[2] Ecologists will point out that these same highland forests are a potential goldmine for these children, who could benefit in a way few can now imagine. There are thousands, if not tens of thousands, of plant species which exist in these forests and in these forests alone, often in only small isolated pockets.[3] Scientific studies of them barely exist, hardly anything is known. If what Thomas Jefferson said – 'The greatest service which can be rendered any country is to add a useful plant to its culture' – is true, then the greatest disservice would be to take one away. It may be a cliché to say the world's rainforests – including Ecuador's cloud forests – could be hiding a cure for cancer, AIDS and other incurable diseases, but it may also be true. Scientists did develop a cure for two forms of cancer from the Madagascan rosy periwinkle.[4] The genetic richness of the cloud forests could one day be as economically beneficial as more traditional sources of income such as minerals or oil.

In the Bellavista reserve, for example, grows the Sangre de Dragón (Dragon's blood) tree, so called because it appears to bleed when you cut the trunk. The sap is used all over Ecuador for treating skin wounds. (As I write, I have Sangre de Dragón rubbed into some painful insect bites – gringos, too, are beginning to discover its benefits!) In any natural medicine shop, Sangre de Dragón is one of the most popular items on sale. So not only is the tree of use to local people and provides some with a source of income, but it might also contain compounds of wider scientific interest. Who knows? We certainly won't if the cloud forests are lost forever.

Contrary to Ipo's pessimism, it may indeed be possible to change local people's attitudes towards these forests. A good day's walk from Bellavista lies a much larger area of cloud forest reserve (over 11,000 acres, as against Bellavista's 700), and here local people are not only clearly aware of the richness of what they possess, but are now being given the chance to exploit this richness without destroying it. Fundación Maquipucuna, which owns the reserve, has been working for nearly a decade to preserve the forest and is now beginning to demonstrate to the local population that they can benefit from projects like ecotourism. Already it employs fifteen people in its tourist lodge and at the end of 1995 it held a training course for people interested in acting as guides to tourists who come to the reserve.

One of the people on that course, Carlos Sevilla, a small farmer from the village of Santa Marianita, epitomises the change in attitude that the foundation is managing to engender, and the huge reserves of environmental zeal that are waiting to be tapped. Well known both within his village and to officials of the foundation, Sevilla was once a prolific hunter of animals from the forest. Now, he says, he is dedicated to trying to supplement his income by using his extensive knowledge of the animals and the forest in ecotourism. Of course, as he points out, there is a high degree of self interest: 'it's easier work.' The fact that he is handicapped by having lost one arm in an accident in a sugar-cane distillery may contribute to his conversion, but his sincerity seems genuine. 'I want to protect the animals for my children, for the people who come after me,' he says.

This concern for the next generation is a recurrent theme among local people. It is not only the influence of the foundation which has shown them certain ways of life are unsustainable; they can see it for themselves. César del Hierro, a village councillor for Marianita, has seen their river decrease in volume during his lifetime. 'When I was little, this river was tremendous, but now it's much smaller. And all this because up in the mountains they have destroyed the forest. This is my main concern.' And Luis Pozo, who has been a park guard ever since Maquipucuna bought the land, says: 'I'm 43. If I plant a tree, am I going to be able to harvest it? No. Nobody knows how long they will live for. The future is for the children.' Pozo admits that when Maquipucuna first arrived with its plans for conservation and other alternative ideas, he was slightly bemused. 'They said, We are going to maintain your forests by setting up a protected forest." I didn't know what they meant.'

Attitudes, then, *can* be changed. This is also apparent in the figure of Jorge Morales, who like Carlos Sevilla, used to make his living from killing Ecuador's big game – particularly the endangered spectacled bear, as well as pumas and jaguars. The fruits of his labours adorn his house – in the shape of their skins – as a macabre reminder of his past. In the hills around his home, the bear was once common, but Morales admits he became concerned by its growing scarcity. With the backing of Maquipucuna, he has devised a plan which will allow him to make money out of the bear alive, not dead. It involves planting a field of corn and building look-out posts at the corners. The bear loves corn and the field would prove an attractive feast, and for anyone – tourists, scientists perhaps – hiding in the posts it would be a rare chance to watch this shy and rare beast in a semi-natural environment.

As of January 1996, the plan was still on the drawing board and whether it will take off is in some doubt. Some conservationists would probably be relieved, pointing out the plan could expose the bear to yet more hunting. Nevetheless, since Morales has now been taken on by the foundation as a park guard, at the very least his new found passion for conservation will be passed on to his children and maybe to the next generation.

The bear – perhaps more than anything else – highlights how neglected Ecuador's cloud forests are, with its existence and fight for survival going largely unnoticed by most Ecuadoreans. In the pre-Columbian era, the indigenous inhabitants of this region revered it as a demi-god, and in one small community in southern Peru, it still is. Now it is largely regarded as a pest because it pilfers corn and other crops. Theoretically, it could act as a talisman to generate interest in conservation among both ecotourists and Ecuadoreans themselves. Maquipucuna employees testify to the excitement created in the area by the release of three, young captive bears into the reserve in December 1995. Now a novel scientific study is being carried out, using a satellite tracking system to discover more about the bear's habits and diet. The point is that if momentum can be given to saving the bear, it would focus attention on the arguably more important issue – its habitat, the cloud forest.

No one is suggesting that ecotourism is the answer to the entire problem, but carried out in a sensitive and sustainable way, it can demonstrate there is an alternative to agriculture and logging. Marcia Valarezo, a former guide who now works for one of Ecuador's most radical, campaigning environmental pressure groups, Acción Ecológica, says ecotourism is a good thing if it allows local people to retain their dignity. In a pamphlet published in 1995, 'Una Propuesta Ecoturística', she expounds her philosophy. 'At the very least, ecotourism should put special care on not negatively influencing the environmental practices of local cultures,' she says. This has been of particular concern in Ecuador's Amazon basin for many years now, where tourism agencies have exploited indigenous communities for their own gains. Andy Drumm, a British environmentalist who is a prominent campaigner for unadulterated ecotourism in Ecuador, has this to say about its possible role in the highlands: 'People here have a weaker link with their physical environment. They have very often come from other areas. Their identification with the land is not as strong as it is for indigenous communities. Ecotourism can play a positive role in strengthening that bond with the land.'

There are other dangers besides the invasion of people's privacy and culture. Ecotourism could, for instance, raise expectations of local people as to how they will benefit, and it could create divisions within the community, with those people who have failed to benefit resentful of those who have. Arcenio Barrera, another employee of Fundación Maquipucuna, says the foundation has annoyed some people as it has gradually bought more land and prohibited hunting on it. 'It's okay for those of us who work, but for those who don't it's not so good,' he says.

Abigail Rome, the reserve's director, says ecotourism will have spin-off opportunities for those not directly employed by the foundation – selling food products and local crafts, for instance. 'Partly it's up to them,' she says. 'We come from a capitalist society and in this world you have to fight to make your money and to succeed.' Of course, Ecuadorean peasants probably know this better than most.

The environmental organization, Fundación Jatún Sacha, now owns several sites in Ecuador, maintained partly through ecotourism. In 1994, it purchased Guandera, 160 hectares of high altitude rainforest in the north of Ecuador, which is under severe pressure from what founder Michael McColm calls 'freelance Colombians cutting down trees for charcoal for the fast food chicken business'. Jatun Sacha has started a programme to take visitors to stay with families and to be shown round the trails by local guides. It is a key part of Jatun Sacha's ecotourism that local people benefit economically from the venture. Jatun Sacha's most recent purchase was 1,000 hectares of low altitude cloud forest on the western coast of Ecuador, where it is encouraging scientific research. The UK organization, Equafor, (Friends of the Equatorial Rain Forests) helped finance the purchase with a donation of £11,000 sterling. Another

British NGO, Rainforest Concern, is spearheading a proposal in conjunction with Fundación Maquipucuna to create a forest corridor ultimately all the way from the area around the Bellavista reserve to the coast. The plan would involve similar work to that which is already carried out by Maquipucuna with local communities, such as reforestation, environmental education and organic agriculture, but would also be expanded to include health and sanitation education and more scientific research. In January 1996, Rainforest Concern was preparing to go to the European Union with its proposal.

Local community initiatives involving ecotourism in the Ecuadorean highlands are notable by their absence. But that does not mean that local people are not involved, nor that they could not one day take greater control over such projects. Ecotourism, by its very nature, is more likely to be an idea conceived by outsiders, particularly in these highlands where true indigenous communities are comparatively rare. The cloud forests are mostly threatened by colonizers, people who have moved in from other regions in the last few decades, people who do not have a long-standing affinity for the region in which they work or for the soil they plough. But it would be wrong of the outside world to assume that these same farmers, given a chance, do not take pride in their environment, and pleasure in the existence of the spectacled bear and mountain toucan. Manuel Ipo, Carlos Sevilla and Jorge Morales will testify to the truth that the cloud forests must be preserved, not just because their children will be economically worse off without them, but because they will be culturally and spiritually poorer as well.

[1] José Vicente Zevallos in William C. Thiesenkusen (ed), *Searching for Agrarian Reform in Latin America*, Unwin Hyman, 1989.

[2] Conservation International, *Status of forest remnants in the Cordillera de la Costa and adjacent areas of SW Ecuador*, October 1992.

[3] Adrian Forsyth and Ken Miyata, *Tropical Nature: Life and Death in the Rain Forests of Central and South America*, p.2.

[4] Edward O. Wilson, *The Diversity of Life*, Penguin 1993, p.271.

4
MODERNIZATION: ENVIRONMENTAL COSTS AND POPULAR RESPONSES

New harvests, old problems: The challenges facing Latin America's agro-export boom

Lori Ann Thrupp

Growing global trade is bringing North American and European shoppers a year-round supply of fresh flowers, fruits and vegetables flown in from Latin America and the Caribbean. In the countries of origin, production of these fashionable 'non-traditional' agro-exports (NTAEs) is booming. Hoping to overcome economic stagnation and add diversity to 'traditional' agro-exports like bananas, coffee and sugar-cane, international financial institutions and national governments have been promoting these products in Latin America for a decade. The NTAE strategy is a key part of trade liberalization and structural adjustment policies. The growth of non-traditional exports has had bittersweet results. They have proved very profitable to foreign investors, some enterprises in the South and to transnational food corporations. They also satisfy the appetites of Northern consumers. On the 'bitter' side, NTAEs have considerable economic, social and environmental costs similar to traditional agro-export sectors, such as high levels of pesticide use. From a macroeconomic point of view, the NTAE boom has been hailed a success. But a closer look reveals symptoms of unsustainability and inequity. Small-scale farmers have been the heaviest losers.

The agro-export legacy

Historically the agrarian structure of Latin America and the Caribbean has been inequitable. Large numbers of people, especially indigenous peoples, have been excluded from the benefits of economic growth and marginalized on unproductive lands. In Central America, for example, large farms over a hundred acres, although representing only seven per cent of total operations, control about 73 per cent of the total agricultural land; by contrast, small farms under five acres, although representing 78 per cent of the total number of farms, occupy only eleven per cent of the total land.[1] In many countries, agro-exports occupy the largest portions of agricultural land. Much of this agribusiness is foreign-owned, especially in Central America. Its main features are large-scale monocultural plantations, high inputs of chemicals, dependency on volatile Northern markets, and the exploitation of both natural resources and low-wage labour. Over time, traditional exports have enjoyed dynamic growth and high profitability but they have also experienced short-lived export 'booms' followed by economic 'busts'.

During the early 1980s, the region suffered a serious economic crisis. Dependency on a small set of traditional exports made Latin American and Caribbean economies vulnerable and unstable. Falling commodity prices, global recession, declining terms of trade and trade protectionism cut into the region's economic earnings. In response to the crisis, most of the region diversified into non-traditional, high-value products. The purpose of NTAE promotion policies and programmes, from the perspectives of international financial institutions and governments, is to generate foreign exchange for repaying the debt, to diversify the economy so as to reduce dependence on low-priced traditional exports, increase agribusiness and export earnings, generate employment, and in general, revitalize economic growth.

Growth of NTAE sector

The expansion of NTAEs was spearheaded by Chile and Mexico whose movement into this sector preceded the regional diversification of the 1980s and 1990s. Between 1962 and 1988, Chile's fruit exports expanded 26-fold, with earnings growing from $19.9 million to $473 million (in constant 1985 US dollars).[2] Following a similar model, Mexico's fresh vegetable sector was developed by US transnational companies in the 1960s. This was followed by the rapid expansion in Mexico in the late 1970s of large agribusinesses in strawberries and tomatoes for export to the US. Recently, the most remarkable agricultural growth in Latin America has been in high-value crops such as flowers, fresh and processed fruits (particularly mangoes, melon, pineapples, passion fruit and berries), and vegetables (such as broccoli, snow peas, asparagus, mini-squash and artichokes). From the mid-to-late 1980s, NTAEs grew by 222 per cent in Chile, 349 per cent in Costa Rica, and by 78 per cent in Guatemala.[3] In Ecuador, the value of NTAEs grew 27-fold, from US$3 million in 1984 to US$75 million in 1994.[4] NTAEs still represent a relatively small proportion of exports compared to traditional exports from Latin America. Indeed it is unlikely they will ever approach the values of coffee, cotton, beef and bananas (nor are they intended to). Nevertheless, the net revenues and returns per acre of NTAEs are remarkably high, especially compared to traditional staple foods. World prices for sorghum, maize and wheat in recent years have ranged from US$75 to US$175 per metric ton, while many NTAE fresh fruit and vegetable products have international prices of US$500 or more per metric ton.[5] The growth of NTAEs in Latin America and the Caribbean since the early 1980s has been strongly supported by international financial institutions, particularly the World Bank and the International Monetary Fund (IMF), and by the US Agency for International Development (USAID). In 1992 USAID spent nearly US$119 million on agribusiness worldwide, the large majority of which was for non-traditional crops.[6] Major trade agreements, including the Caribbean Basin Initiative, NAFTA and GATT, have also spurred export growth, creating incentives for trade liberalization.

The winners

The main beneficiaries of NTAE growth are large companies – both transnational corporations and large national and foreign investors. Many crops are grown in high-tech conditions and are capital-intensive, which usually limits them to wealthy enterprises. In Central America, multinational corporations account for approximately 25 per cent of the total production of NTAEs.[7] Transnational business is strongest in fruits and vegetables. For example, Del Monte in Costa Rica and Dole in Honduras market almost all pineapples exported from these countries.[8] Besides the transnational corporations, another important category of NTAE enterprises are large and medium-sized businesses, with foreign investors playing a highly prominent role. In Costa Rica, they dominate the production of flowers, ornamental plants, citrus fruits and macadamia nuts. Of the fourteen largest flower growers, all but two are foreign.[9]

In some cases, small-scale, resource-poor producers have benefited considerably from NTAEs, especially when organized in cooperatives or groups. Guatemala provides a unique illustration of small farmers' substantial involvement in NTAEs, especially in vegetable production. Ninety per cent of Guatemalan snow peas are grown by relatively poor farmers on very small farms – typically less than two acres.[10] Some of these farmers are organized into cooperatives, which help facilitate marketing and technical services. The most renowned cooperative involved in NTAEs in Guatemala is Cuatro Pinos, which has 1,700 members and developed a thriving enterprise during the 1980s. In Bolivia, too, small-scale farmers have been involved in NTAE production through cooperatives and traditional associations. Many smallholders produce *quinoa*, a traditional protein-rich grain that indigenous people have cultivated for centuries, using little or no chemical inputs. This grain has recently been rediscovered by natural food fans in North America and Europe, offering new export opportunities for Andean indigenous producers. Bolivian smallholders are also producing organic chocolate for export. The competitiveness of small farmers in these crops is largely due to relatively low labour requirements, low land rent, and the use of family labour. In the 1990s these groups have suffered from declining prices and growing national and international competition. Nevertheless, such producers are more likely to achieve success when they are organized into associations.

The losers

However, on the whole, resource-poor farmers have great difficulties competing in the NTAE market. In nearly all countries, they lack access to the credit, capital and information required for success in the business. They also tend to be unfamiliar with the crops and the production technology and have little access to technical assistance services. Gaining entry into the market is particularly difficult for poor farmers in countries like Chile, Ecuador, Paraguay and Costa Rica, where NTAE programmes have given very little attention to their concerns. NTAE business associations are usually beyond the reach of poorer producers, since they tend to have high membership fees. Nor can poor farmers afford the expensive transport systems and storage facilities required for rapid and effective marketing. Even if they do get involved in NTAE production, they are likely to find themselves squeezed out of the competition by larger producers, as in the case of the Chilean fruit industry. Even in Guatemala, many small farmers have been unable to maintain a profitable business in NTAEs when faced with falling prices and the superior marketing resources of larger producers. NTAE farmers also suffer from a lack of diversification. At the national level, the diversity of NTAE products has increased significantly in all countries, but farmers in the NTAE sector usually plant entire farms with a single monoculture. For small farmers, this generally means converting from mixed farms growing a variety of subsistence crops into farms planted with a single export commodity. Over time, this uniform conversion can bring various economic risks and other disadvantages. By contrast, maintaining diversity of crops within individual farms can help ease the risk factor, reduce vulnerability to fluctuating prices, reduce pests and diseases, and holds other agro-ecological advantages for soils and resource nutrients.

Throughout Latin America and the Caribbean, a large proportion of workers in non-traditional agro-export sectors are women, in both production and processing.[11] This increasing feminization of the rural wage labour force is a ubiquitous change that has accompanied the globalization of food systems. Traditionally women have been actively involved in subsistence farming but the agro-export industry has increased the amount of waged work available to women. In Ecuador in 1991, for example, an estimated 69 per cent of workers in NTAE production were female.[12] In many cases, an important reason why managers prefer women is that they are paid lower wages than men for equivalent work; they also work longer hours (often without over-time pay), are rarely promoted and are poorly protected by labour laws (see Sarah Stewart's article).

NTAEs and the environment: more pesticides, more problems

The production of NTAEs is characterized by the heavy use of chemical inputs and pesticides. Studies have shown consistently that all kinds of pesticides, including fungicides, insecticides, nematicides and herbicides, are used more intensively for most high-value NTAEs than for other crops.[13] Doses of pesticide applications per unit of land in NTAEs exceed those used in subsistence crops and for crops sold in local markets and are similar or even greater per hectare than in many of the traditional export crops such as coffee and sugar-cane. Within the NTAE sector, the amounts of pesticides applied are particularly high for perishable fresh fruits, vegetables and flowers.

Although pesticides can bring short-term benefits in controlling pests, heavy pesticide use has several adverse effects. Firstly, pesticides are expensive. On Ecuador's rose plantations, in one year an estimated average of 35 per cent of operating costs was spent on agrochemicals – of which 85 per cent was for insecticides and fungicides.[14] Studies in Guatemala have also documented high direct costs for pesticides. The Consortium for International Crop Protection found in the late 1980s that pesticide costs for melons had reached US$735-2,206 per hectare and for snow peas had exceeded US$2,206 per hectare.[15] According to the study, pesticide purchases, application and technical assistance costs for NTAE vegetables accounted for 22.5 per cent of total production costs. A recent study of pesticide use for snow peas in the Guatemalan highlands indicates that pesticides represent about 30-35 per cent of material production costs.[16] Another recent study showed that snow peas generate higher pesticide costs per hectare than either cotton or bananas, which formerly used the highest levels of pesticides per unit of land.[17]

When pesticides are applied excessively or too close to harvest time, the residues accumulate in foods at levels that exceed the tolerance standards established by the governments of importing countries. Pesticide-residue violations and detentions have been a major problem for Latin America and Caribbean NTAE exports to the United States. Detentions have occurred more than 14,000 times in the last ten years at US entry ports, totalling approximately $95 million in imported produce.[18] Violations have resulted in millions of dollars in losses for the exporters and producers.[19] The most serious and frequent residue-detention problems in the region have occurred with products from Guatemala and Mexico. In Guatemalan NTAEs during the late 1980s, detention rates reached 27.3 per cent of the total shipments.[20] Between 1990 and 1994, Guatemala's exports were detained 3,081 times because of residue violations, resulting in total losses of about US$17.7 million.[21] Mexican export crops were detained 6,223 times in the 1980s, resulting in losses of US$49.5 million, and 1,391 times in the 1990s, with losses valued at US$5.9 million. The major residue problems here were found in peppers, strawberries, and mangoes, and the principal pesticide involved was methamidophos, a toxic product with high health risks.[22] Residues have also been high in NTAEs from the Dominican Republic. In 1987 and 1988, shipments from the Dominican Republic with illegal residues far exceeded those from other countries; 12.2 per cent of the samples violated US government standards.[23] The problems were particularly serious in oriental vegetables: in 1988, several growers reported residue-related losses of hundreds of thousands of dollars in a single month. The FDA subsequently imposed automatic detentions on five vegetables, requiring exporters to pay an additional charge of US$400 for each cargo container tested in the United States.[24]

Another negative effect of continual use of pesticides is pest resistance. Through genetic selection, pests evolve to tolerate the effects of pesticides. As pesticides become ineffective, high economic losses ensue, and farmers become trapped in a cycle of increasing pesticide use in an attempt to regain control. Pesticide resistance is sometimes accompanied by the death of natural pest enemies, leading to outbreaks of secondary pests. The resulting 'pesticide treadmill' has already affected many traditional agro-export crops in Latin America, leading to major crop losses. The same treadmill is now being recreated in the NTAE sector.[25] The heavy use of pesticides in NTAEs has contributed to a resurgence of whiteflies in Latin America and the Caribbean, reaching crisis proportions in some areas.[26] Pesticide-resistant whiteflies are also transmitting viruses that have seriously damaged NTAE crops in Chile, Brazil and Argentina. More than one million hectares of cropland in South America have been abandoned due to whitefly-transmitted viruses.[27] In Honduras, outbreaks of resistant whiteflies and leaf miners reached a crisis in the late 1980s, reducing melon harvests by 45-56 per cent in the Choluteca region.[28] Some melon producers lost entire crops. Such losses particularly devastated small producers, who were often forced out of competition. Specialists identified frequent pesticide applications and the promotional tactics of pesticide salesmen as the main sources of the problems.[29] Pesticide use also puts workers' health at risk. Increasing numbers of people are being exposed and impaired, and more are suffering acute poisonings and chronic damage. Most of the victims are agricultural workers – the poorest of those involved in NTAE production. In a recent survey of workers in Ecuador's NTAE sector, 62 per cent of respondents said that they had suffered health disorders from exposure to pesticides while at work.[30] Health hazards are particularly serious in flower production, especially in Colombia and Ecuador, partly because toxic nematicides such as aldicarb and fenamifos are heavily used. Although aldicarb was banned in Ecuador in late 1991, it was still used as late as 1993 because flower growers value its efficacy more than worker safety. To compound the problem, flower managers commonly allow unprotected workers to continue working alongside workers applying chemicals.[31]

Inevitably, the growth of NTAEs has changed the use of natural resources, particularly land, vegetation and water. Deforestation for NTAEs appears to be less than the heavy forest conversion for traditional export crops but preliminary appraisals suggest that significant areas of forest cover have been cleared for NTAEs in certain areas, such as the Central Valley in Chile, and the north of Costa Rica where approximately 3,000 hectares were cleared for citrus plantations. Perhaps more significant than forest removal are the agro-ecological changes provoked by attempts to standardize products so that they conform to a uniform foreign variety, as required by Northern markets. This change tends to undermine the genetic diversity of indigenous varieties and makes the crops more vulnerable to pests and diseases.

The risk factor

As already mentioned, NTAEs tend to be grown in monocultural farming systems. This dependency on one crop or species not only leaves farmers vulnerable to price fluctuations, but also increases the potential for major losses from poor weather conditions. Current services for NTAEs are largely dependent on foreign aid, but this support is being cut back. Even though some market studies suggest that Northern demand for NTAEs will increase, the market may not expand enough to absorb the growing supply. Since many of these products are 'trendy' luxury foods susceptible to instabilities, economic recession in the North, as well as changes in consumers' tastes, can reduce the demand and thus curtail market opportunities. Given these uncertainties, some analysts predict that the present fervour surrounding NTAEs will soon fade. Last, but certainly not least, one of the crucial issues noted by several analysts and numerous farmers is that non-traditional agro-export growth and the associated change in land use reduce food availability locally, thus threatening food security, both at local and national levels. In nearly all countries, fiscal policy changes to stimulate NTAE growth have reduced funding for local food production. Likewise, analysts have noted that the increasing investments by international institutions in export-directed development have come at the expense of attention to domestic food needs.

Moving towards sustainability

If we take into account the social and environmental costs of NTAEs, the supposed 'success' of the nontraditional agro-export development strategy is doubtful. The expansion of this new sector is repeating the same patterns and risks which were characteristic of past agro-export booms. As 'free trade' measures become increasingly influential, these risks will be aggravated. Meanwhile, local food production in Latin America is stagnating, and hunger and insecurity among the majority of the region's rural people will continue to grow. Changes in agro-export growth policies and practices are needed at all levels to prevent problems and to increase the benefits in this sector. Environmental incentives and regulations and the social needs of the poor should be fully integrated into agricultural and economic policies and trade agreements. Market forces and short-term, unfettered agro-export growth will not guarantee sustainable development. Programmes must be redesigned to help expand equitable opportunities for the poor and to eliminate policies that degrade the environment, such as incentives for pesticide use. Producers, non-governmental organizations and agrochemical suppliers as well as government agencies and development organizations need to develop sustainable production methods, such as the promotion of *indigenous* crops with a high market potential. They should also help strengthen local farmer organizations and improve technical services so that the opportunities and bargaining power of the rural poor are increased. More fundamentally, the underlying problems of inequitable agrarian structures in Latin America and the financial pressures from international financial institutions need to be addressed. These kinds of changes are needed not only for social and ethical reasons, but also for the good of national economies and political stability in the region.

*Revised and updated version of an article of the same title which first appeared in *NACLA Report on the Americas*, Vol XXVIII No 3, New York Nov/Dec 1994. For additional information, please see *Bittersweet Harvests for Global Supermarkets: Challenges in Latin America's Agricultural Export Boom,* by Lori Ann Thrupp (1995). Available for US$19.95 (plus $3.50 shipping) from World Resources Institute Publications, P.O. Box 4852, Hampden Station, Baltimore, MD 21211; phone (800) 822-0504.

[1] *VIII Compendio Estadístico para Centro América* SIECA, 1978; cited in T. Barry, *Roots of Rebellion: Land and Hunger in Central America*, South End Press, Boston, 1987.
[2] P. Tabora, 'Central America and South America's Pacific Rim Countries: Experience with Export Diversification,' in S. Barghouti, L. Garbus and D. Umali (eds), *Trends in Agricultural Diversification*, World Bank Technical Paper #180, World Bank, Washington DC, 1993.
[3] Cited in M. Carter, B. Barham, D. Mesbah and D. Stanley, 'Agro-exports and the Rural Resource Poor in Latin America: Policy Options for Achieving Broadly-based Growth,' unpublished draft paper, Univ. of Wisconsin, Madison, 1993; based on USDA statistics.
[4] K. Swanberg, 'Agribusiness Assessment: NTAEs in Ecuador,' draft paper, USAID, Center for Development Information and Evaluation, Washington DC, 1994.
[5] S. Jaffee, 'Exporting High-Value Food Commodities: Success Stories from Developing Countries,' World Bank Discussion Paper #198, World Bank, Washington DC, 1993.
[6] Cited in M. Lycette, 'Women, Poverty and the Role of USAID,' in *Poverty Focus*, Bulletin, #2, Overseas Development Council, Washington DC 1994.
[7] J. Collins, 'Nontraditional Agro-exports, Basic Food Crops, and Small Farmers in Central America,' unpublished paper, Interamerican Foundation, Arlington, VA, 1992.
[8] Ibid.
[9] D. Kaimowitz, *El Apoyo Necesario para Promover Exportaciones Agrícolas No Tradicionales en América Central* #30, Interamerican Institute for Cooperation on Agriculture, San José, Costa Rica.
[10] J. Fox, K. Swanberg and T. Mehan, 'Agribusiness Assessment: The Case of Guatemala,' draft paper, USAID, Washington DC, 1994.
[11]. Lucía Salamea, A. Mauro, M. Alameida and M. Yepez, 'Rol e Impacto en Mujeres Trabajadoras en Cultívos No Tradicionales para la Exportación en Ecuador', CEPLAES, Quito, 1993; and Rae Blumberg, 'Gender and Ecuador's New Export Sectors,' draft paper, GENESYS Project, Report for USAID, Washington DC, 1992.
[12] Proyecto para Exportación Agrícola No Tradicional (PROEXANT), 'Reporte Anual,' unpublished report, PROEANT, Quito, Ecuador, 1993.
[13] D. Murray and P. Hoppin, 'Recurring Contradictions in Agrarian Development: Pesticide Problems in the Caribbean Basin in Nontraditional Agriculture,' *World Development*, 1992, vol. 20, no.4:603; I. Merwin and M. Pritts, 'Are Modern Fruit Production Systems Sustainable?', *HortTechnology*, April-June 1993, vol.3, no.2:131. Also AGROSTAT data from FAO, Rome, Italy.
[14] W.F. Waters, 'Restructuring of Ecuadorean Agriculture and the Development of Nontraditional Exports: Evidence from the Cut Flower Industry,' Quito: Universidad San Francisco de Quito, 1992. Paper presented at the 55th Meeting of the Rural Sociological Society, University Park. PA.
[15] CICP, 'Environmental Assessment of the Highland Agricultural Development Project,' Project Amendment, Phase 2, College Park, Md.: Consortium for International Crop Protection, 1988.
[16] R.W. Fisher, R. Cáceres, E. Cáceres and D. Dardón, 'Informe Final: Evaluación del Manejo de Plagas y Plaguicidas en Arveja China del Altiplano de Guatemala,' unpublished final report on study by Centro Mesoamericano de Tecnología Apropiada, El Instituto de Ciencia y Tecnología Agrícola, and World Resources Institute, Guatemala City, Guatemala, April 1994.
[17] D. Murray, *Cultivating Crisis: The Human Costs of Pesticides in Central America*, University of Texas Press, Austin, 1994.
[18] US Food and Drug Administration, primary unpublished data, Washington DC, 1983-1994 (analyzed and compiled by World Resources Institute).

[19] Ibid.

[20] D. Murray and P. Hoppin, op.cit.

[21] Calculation of WRI, based on analysis of US Food and Drug Administration primary data.

[22] Ibid.

[23] D. Murray 1994, op.cit.

[24] D. Murray and P. Hoppin, op.cit.

[25] L.A. Thrupp, 'Entrapment and Escape from Fruitless Insecticide Use: Lessons from the Banana Sector in Costa Rica,' *International Journal of Environmental Studies*, 1990, vol 36:173-189; D. Murray, op.cit.; D. Bull, *A Growing Problem: Pesticides and the Third World Poor*, Oxfam Books, Oxford, 1982.

[26] F. Morales, 'Controlling Plant Viruses in a Changing Agricultural Environment,' unpublished article, Centro Internacional de Agricultura Tropical, Cali, Colombia. Also see D. Murray, 1994, op.cit.

[27] F. Morales, op.cit.

[28] D. Murray, 1994, op.cit.

[29] Ibid.

[30] PROEXANT/University of San Francisco at Quito, unpublished results of 1992 survey of NTAE producers. See also Jorge Rodríguez in W. Waters (ed), *Desafíos de los Cultivos No Tradicionales en Ecuador*, USFQ, Quito, 1992.

[31] R. Blumberg, 'Gender and Ecuador's New Export Sectors,' op. cit.

The price of a perfect flower: environmental destruction and health hazards in the Colombian flower industry

Sarah Stewart

Carmenza bends tenderly over her baby son, adjusting the shawl in which he is carefully swaddled. He is so young that he does not even have a name, but already he's been admitted to hospital with bronchial problems. 'The doctor said it was because of the pesticides I breathed in while I was pregnant,' Carmenza says.[1] She is one of 72,000 workers, most of them women, who work in Colombia's flower industry. For the price of a small bunch of carnations in a typical London florist, she works eight to twelve hours a day – perhaps more during the peak season before Valentine's Day when 14 million roses are jetted off to the United States alone.[2] She is proud of the beautiful blossoms she grows. But she is worried about her health and that of her child.

Fainting, dizziness, skin irritations, respiratory ailments, neurological problems, premature births – these are some of the symptons which flower employees and doctors say are due to pesticides at work. At one regional hospital, up to five people a day arrive, acutely poisoned, with another twenty showing signs of chronic effects.

Pesticide poisoning: is this the price of Colombia's perfect flowers? No one knows what the exact damage is. There are no scientific studies of health or of the impact on the environment of the Bogotá savannah where the industry is based. Indeed, conclusive causal studies would be difficult to carry out, so intertwined are poverty, malnutrition, and pollution in the savannah. It is evident that many flower companies are responsible employers. But from anecdotal evidence, many are not. In town after town, there are worrying reports of illness and environmental damage and of fears that the flower industry may be exacting too high a price.

Flowers are a Colombian success story, a triumph of Third World, agro-exports and sophisticated marketing. In 1965 a handful of companies set up shop in Madrid, an agricultural town in the western savannah. Thirty years later, there are over 450 companies nationwide, exporting over 3.5 billion flowers a year. Roses and carnations, alstroemeria and chrysanthemums have seized foreign markets with remarkable success. Over half of all roses and 95 per cent of all single-stem carnations sold in the US come from Colombia, as do over half the carnations sold in the UK – roughly 33 million a year.[3] Under the threat of competition from low-wage havens such as Kenya, Morocco, Zimbabwe and Ecuador, the pressure is on to keep labour cheap and hours long.[4]

The flower industry is the main employer in 25 out of 27 towns in the Bogotá savannah, supporting 120,000 households nationwide. Political power in the area lies with flowers, too. The mayor of Funza, for instance, is a grower himself. Taxes are low; despite the sophistication of the enterprises, it is categorized as agriculture and pays less than industry. When the town of Sopó tried to impose a license fee for operating within the municipality, it was told that municipalities had no right to impose standards.[5]

Pesticides and the perfect flower

In the plastic-sheeted greenhouses which line the broad, fertile plains of the savannah, dozens of pesticides are routinely sprayed to give the flowers the perfect look they need to enter US, European or Japanese markets. A single blemish, the mark of an insect, or a curling leaf, will stop a shipment, so strict are international standards. Between 70 and 180 kilos of active ingredients (as much as 333.6 kilos of pesticides) are used per hectare of flowers each year.[6] Up to twenty per cent of these pesticides may have been banned or not registered in Europe or the US.[7] [8] Some are illegal under Colombian law. But legal or illegal, pesticides are toxic by design. If they are not used carefully, they can cost lives.

'When a person is exposed repeatedly to [organophosphate] pesticides,' explains a doctor, 'alterations occur in the nervous system and then the whole body.' Multiple sclerosis, motor neurone disease, and Parkinson's disease may be the grim results, say some scientists.[9] Developed in the 1930s as nerve gases, these chemicals can also produce cramping, loss of vision and muscle strength, tremors, dizziness, partial paralysis, optic neuritis, and depression of the cardio-repiratory system.[10]

But organophosphates are only one type of pesticide used in the industry. There is also, for instance, the fumigant methyl bromide, categorized as acutely toxic by the World Health Organization. Overexposure can cause pulmonary edema. Etilio-tio-urea, a fungicide ingredient, is carcinogenic and may cause birth effects. Exposure to the insecticide, lambda cyalotrina, WHO category II (moderately hazardous), may have sent thirteen people to casualty at San Pedro Claver clinic recently, all suffering from muscular weakness, tingling in limbs and loss of strength.[11] But it is the long-term risks that are especially worrying: miscarriages, premature births, hyperactivity of the bronchial system, and respiratory and neurological problems. Lack of water and malnutrition – both endemic – may increase toxicity. Captan, for instance, the most commonly used fungicide in flowers, is said by the US Environmental Protection Agency to be a probable cancer-inducing agent. When fed to protein-deficient rats, its impact grew 2,000-fold.[12] Young people are especially vulnerable.[13]

In early 1994 the Colombian Ministry of Health banned some organochlorine pesticides and three foreign companies – Ciba Geigy, Hoechst and Schering – have voluntarily withdrawn a number of dangerous pesticides. But who is to guarantee that such bans will stick, or that Colombian health and environmental law will be enforced? Government presence is minimal and that which exists is often disregarded. Despite the efforts of the Association of Colombian Flower Exporters (Asocolflores) to educate its members, adherence is a matter of choice. Many employers follow the law, but, as one Ministry of Health official commented, 'compliance is up to the company.'

Exposure to pesticides is highest in the greenhouses. In Dutch flower greenhouses, heat and lack of circulation accentuated the toxic effects by sixty times the acceptable levels, according to a study published by the Dutch government.[14 15] Sprayers are also highly exposed. The government-recommended training time for sprayers is 60 hours. But fewer hours seems the rule, along with reports of torn masks and inadequate protective clothing. US EPA guidelines recommend staying outside greenhouses for 12 to 48 hours after spraying. But in some companies, if only one flower bed in a greenhouse is being sprayed, the women are not removed at all. 'They send us back to the greenhouse within an hour of spraying,' says one woman, 'while the flowers are still wet. I know it's a threat to our health.'

Almost all flower companies have resident doctors or register their employees with a nearby social security clinic. Testing for cholinesterase levels – the main existing test for pesticide contamination – is carried out routinely, particularly with fumigators. Many have occupational health committees. Asocolflores distributes comic books on health issues to workers and a monthly bulletin advising employers on international standards and appropriate pesticide use. Yet many workers are 'completely unprotected'.[16] Doctors and workers report that the results of the cholinesterase blood tests are rarely given to the workers. A doctor says: 'They do tests just thorough enough to make sure the worker can carry on producing.'

Work in the flower industry is not only hazardous, but hard. The working day is long – 8 to 12 hours, with perhaps only 15 minutes for lunch – and the work is heavy and repetitive. When they take off the rubber bands holding the flower buds, many women's fingers bleed because of the speed with which they must do the work. The pay is just over the minimum wage of $118 (£75) per month, plus a thirteenth month salary and overtime. Working in the flower industry used to be the preferable and more sociable option compared to domestic service. But that security is now evaporating. Under Law 50, employers can hire workers on six or even three month contracts and dismiss them without cause. Trade unions, already weak, have been further undermined. Elvira, who works in a rose greenhouse, complains: 'I used to have 20 rose beds. Now I've got 25 to 30 beds for the same pay. If you don't accept the new quota, they sack you.'

Workers' resistance

Women and men in the flower trade have fought to have labour and environmental laws enforced and for the right, for instance, of pregnant women to be shifted to safer and less strenuous jobs. Many of them have fought for unions. This has required considerable courage. Workers have been sacked for having spouses who were trade unionists, for chatting to colleagues, for 'taking too long to take out the rubbish' and for any number of alleged infractions of the rules. An employers' blacklist is said to weed out troublemakers. 'If you demand your rights too much, they fire you,' says Wilfer, a worker in his twenties. Another worker, Amanda, was sacked for advising a colleague to go to the Ministry of Employment when denied maternity rights.

Workers are tenacious about knowing their rights and getting them enforced. But in many cases the odds are against them. Although there is no overt repression in the Bogotá savannah as there is in the rest of Colombia – a country with one of the world's highest records for murdered trade unionists – there is a climate of fear. Workers are too vulnerable, and dismissals under Law 50 too easy for trade unionists to survive with ease.

Flower children

Health problems in the savannah are difficult to disentangle from the more general ones of malnutrition, poverty and pollution. It is particularly difficult to isolate cause and effect with children, who are first to be vulnerable to malnutrition and general ailments. Anecdotal evidence, however, suggests that there may be serious effects of pesticides on children of flower workers. These include:

1) high levels of miscarriage and premature births among women (this is also related to continual bending and kneeling for up to 12 hours by pregnant women and to poor nutrition)

2) high levels of respiratory problems among infants and toddlers

3) high incidence of rashes and other skin irritations

4) possible pesticide poisoning of babies and small children through breastmilk or skin contact.

Amelia's baby, for instance, was five weeks old when he took ill. She rushed him to hospital, gasping for air and unable to breathe, where he was resuscitated. Her breastmilk, the doctor said, was contaminated; she had poisoned her own child.[17 18]

But poverty takes its toll, too. Some families live in one-room bedsits without ventilation or adequate sanitation, the cold rising from the concrete floors and walls, a kitchen shared between perhaps four families; dirt floors turn to mud during rains. Food prices are high: about half the British equivalent, more for chicken or other luxuries. Long hours at work make it difficult to give children the care and attention they need; many children are left alone at home after school.

'The entire savannah is contaminated'

Alongside a flower farm in the outskirts of Madrid, a girl walks to the neighbourhood well which is unlined by concrete and simply dug out in the dirt. She tugs off the wooden lid and pulls up the bucket: the water is putrid, smelling terribly and coloured an alarming yellow. Larvae squirm in the foul water. 'We've had to use this for drinking and cooking,' says her father. The mayor assured people that the water was safe from the effects of leaching. But the family has its doubts. Since a community strike two weeks earlier, water has been delivered by truck daily. But they still have to rely on well water for bathing and sometimes to wash clothes and dishes.

Water shortages are common. One large flower company may consume as much water as a rural community of 20,000. In areas of heavy cultivation, flower companies may consume three quarters of all water used. In this area once so rich in water resources, water table levels have dropped to 25 metres.[19] Near the town of Madrid, a once large river meanders slowly. It is smelly, polluted and filled with algae, a sign of the depletion of surface water. The result is rationing: taps which work every other day, or water brought in by truck. People may have a mere seven litres a day – for bathing, washing up, laundry and drinking. Up until 1992 (when a severe drought exacerbated the problem) water shortages provoked protests across the savannah. Today, Madrid, Funza and Mosquera have made a water supply deal with the Bogotá water company – but only at a cost increase of an average $18 per month per family.

There is, moreover, a 'systematic lack' of water treatment.[20] Some flower companies do not treat their waste water at all; it is simply left to run off into rivers or irrigation ditches, or to filter into the ground where it leaches into ground water. Asocolflores monitors local water and encourages the rational use of water resources, but no one has determined the level of toxicity in savannah water supplies. And some chemicals will not go away. Although most were banned in 1994, organochlorines are highly persistent, with a half-life of ten years. These pesticides are retained in the soil, grass, water, the food chain and fatty tissue, including that of cows' and mothers' milk. 'Forty years will pass before these organochlorines are out of the system,' says a Ministry of Health specialist. Even flower stems can be dangerous. Some companies sell them to local farmers to feed to livestock, transferring pesticides from the stems into the milk and meat which are then locally consumed. But even though people might be aware of possible contamination, the incentive is to buy local milk – at less than half the price of milk trucked in from Bogotá.

Entering the Pablo Neruda neighbourhood of Sibaté on a bumpy dirt road, the view from above is of a tiny cluster of houses entirely surrounded by flower greenhouses. When the flower companies spray, as they do several times a week, the residue floats into homes, schools, shops, and onto the laundry – sheets, children's pyjamas, sweaters, jeans – hanging on the chain-link fence separating the houses from the farms. 'Smoke covers the *barrio* when they [the greenhouses] burn the fires [of waste]. We can smell the chemicals when they spray. Dust gets on people, and the dust is covered in chemicals,' says a doctor in the health clinic. Health problems are chronic and widespread.

In 1993, residents of Sibaté passed a resolution to force the companies to move their greenhouses away from residential areas. New companies were told to set up 500 metres away from homes and existing farms were given over a year to relocate. But in the end, despite the intervention of the federal government's Attorney General for Environmental and Agrarian Affairs, nothing was done. The companies argued that a move would prove financially onerous and that jobs would be lost; Pablo Neruda remains surrounded by greenhouses.[21]

Good employers: 'We guarantee the welfare of our people'

Is it necessary to poison the environment and the flower workforce to make a profit? Companies such as Rosex, SA, show that it is not. The setting itself is tranquil: long, graceful rows of roses line the greenhouses; the classification room is clean and airy; a donkey trundles along the path bearing a huge load of long-stemmed yellow roses. Two hundred and fifty people work here. Each receives a medical check-up and is affiliated to social security on entry into the company. Fumigators are given blood tests monthly for cholinesterase levels; women are checked three times a year. Even management is tested regularly. Sprayers are also rotated to reduce exposure and are given full protective clothing.

There are few of the petty restrictions (certain hours for going to the toilet, prohibitions on talking to colleagues, etc) which appear to operate in so many companies. Women in the classification room are given different tasks every few weeks to avoid repetitive strain. And most workers, at least as of early 1994, are hired on indefinite contract. 'It's so much more expensive to be teaching new people every two to six months what to do,' says Guillermo León. 'People have to know their roses... He or she has to know how to apply the scissors. So it's important to offer workers stability.'

Like Rosex, Asocolflores argues that there is an economic logic to improving labour and environmental standards. It gives as an example integrated pest management, which can cut pesticide costs dramatically, or the introduction of new forms of composting carnation stems and leaves. It is also involved in research on the use of natural enemies to control insect pests, water quality in the savannah, and reducing pesticide use in greenhouses, and in promoting recycling of glass and plastic pesticide containers and the use of water filters for waste water. Symposiums, newsletters, and meetings help promote this new push for 'eco-efficiency'.[22]

But reports of improvements are greeted with scepticism.[23] Increasing quotas, pressure to produce, the constant threat of sackings, job insecurity, the flagrant lack of local government control over flower company practices all point to the continuing violation of Colombian law and of workers' and community rights. The Colombian government estimates that the top ten per cent of flower companies follow Colombian law. Yet this still leaves hundreds of companies, and an unknown number of subcontractors, which may *not* be abiding by Colombian law. Nor does the government seem to care, despite excellent legislation and the existence of committed people in the ministries. In the words of one researcher, 'it is not taking on board... what can well be called the destruction of the savannah of Bogotá and of the flower workers.'[24]

A fair flower

So notorious are the conditions in the savannah that the industry has become the focus of a European campaign to improve working and environmental standards. Begun in Germany, Austria and Switzerland in the early 1990s, and taken up in Britain in 1994, the campaign has tried to publicise the worst excesses of the industry. Working closely with Colombian NGOs, the campaign is investigating the prospects of a 'seal of approval' for flowers imported by companies which respect the rights of their workers and the communities in which they operate. By exerting pressure from consumers internationally, the campaign hopes to help raise the standards of the entire industry.

'We like producing beautiful flowers,' one woman said. 'We put a lot of care into them. But it's distressing that we're being treated so badly. We don't think people should stop buying flowers. But they should help us to improve conditions.'

[1] Unless otherwise indicated, all interviews were conducted during a research trip undertaken for the development agency Christian Aid during February/March 1994. Names of most workers have been changed to protect their anonymity.
[2] Calculation based on information in Niala Majaraj and Gaston Dorren, *The Game of the Rose: The Third World in the Global Flower Trade*, Utrecht, International Books, 1995, p.42.
[3] The US market dominates, with three-quarters of 1994 exports of $427 million; Europe takes around 14 per cent, with Britain the single most important European market.
[4] International financial organizations have been keen to encourage Third World countries to diversify into flower exports as part of structural adjustment packages, with the likely result being oversupply on the world market. See *The Game of the Rose* op cit and Niala Maharaj, 'Tropical Flowers: Miracle Crop with a Flaw', *Development and Cooperation*, January 1995, Debbie Hamrick, 'For Colombia, the Business of Flowers is Change', *Floraculture International*, January/February 1994.
[5] 'Las Arcas no se Llenan', Grupo Cactus, *Cactus*.
[6] *Game of the Rose*, p.58; see also Dr Uwe Meier 'El Cultivo de Flores en Colombia', study commissioned by Asocolflores, December 2, 1992.
[7] Bread for the World and Pesticides Action Network, 'Pesticides in the Colombian Flower Industry'; data analysed by the Pesticides Trust, London.
[8] See also 'Trabajo de Niños y Jóvenes en la Florcultura en el Municipio de Madrid, Cundinamarca', Universidad Nacional de Colombia, Centro de Estudios Sociales (CES), Santafé de Bogotá, 1995, p.30.
[9] Mark Redman, 'Organophosphates: The Pyramid of Exposure', *Living Earth*, November 1993, and Mark Purdey 'Anecdote and Orthodoxy: Degenerative Nervous Diseases and Chemical Pollution', *The Ecologist*, Vol 24, No 3, May/June 1994.
[10] Samuel Henao, Jacobo Finkelman, Lilia Albert and Henk W. de Koning, *Pesticides and Health in the Americas*, Washington DC, Pan American Health Organization, February 1993, p.3.
[11] 'Trabajo de Niños y Jóvenes', Universidad Nacional de Colombia, p.32.
[12] 'Pesticides and Health in the Americas', PAHO, p.2.
[13] Children are especially vulnerable because they are still growing. Young people between 14 and 18 are commonly employed during school holidays or as school leavers. See 'El Trabajo de Niños y Jóvenes', op cit pp.6 and 30-34.
[14] Grupo Cactus, *Cactus*, boletín informativo, September 1995.
[15] *Game of the Rose*, op cit p.101
[16] Testimony by Dr Ernesto Jairo Luna, European Parliamentary hearings on Colombian flowers, Brussels, April 11-12, 1994.
[17] Interview, July 1992.
[18] In a study conducted by the Finnish Institute of Occupational Health, women agricultural workers exposed to pesticides in the first three months of pregnancy were twice as likely to give birth to children with cleft palates or defects in the central nervous system (see Grupo Cactus, *Cactus*, September 1995).
[19] Dr Uwe Meier,'El Cultivo de Flores', op cit p.5.
[20] Dr Ernesto Jairo Luna, testimony at European Parliament hearings, Brussels, April 11-12, 1994.
[21] Grupo Cactus, *Cactus*, September 1995.
[22] Asocolflores, 'Reseña Asociación Colombiana de Exportadores de Flores', 1995.
[23] Mary Matheson, *The Scotsman* 14 February 1996.
[24] 'Trabajo de Niños y Jóvenes', op cit, p.36.

David vs. Goliath: fishermen conflicts with mariculturalists in Honduras

Denise Stanley

What can a disparate, powerless group of people do to reverse environmental degradation? And how are current trends in the exploitation of natural resources affecting the achievements and sustainability of environmental activism? One of the most contentious natural resource conflicts in Latin America today is the expansion of shrimp farming mariculture. The following case study of shrimp farming in southern Honduras sheds new light on these questions. Although mariculture industries have been developed in Guatemala and Nicaragua, Honduras is the Central American leader in the new shrimp export boom. This is partly because the Honduran state has given fiscal incentives to export producers in its drive to open up the Honduran economy since 1985.[1] Shrimp and other seafood products alone are now the country's third largest export in dollar value.[2] Aquaculture has been described by some as an effective way of privatizing fisheries and avoiding their overexploitation.[3] Others have questioned the long-term impact of the mariculture industry, while a highly polarized debate has emerged at the local level regarding land access issues and environmental degradation. The root of discontent stems from the 'enclosure of the commons' and the perception that the shrimp industry is reducing fish stocks on which the zone's resident fishermen depend.[4] Similar to other environmental activists elsewhere in the Third World,[5] the rural poor of southern Honduras have resisted the commercialization of local space because of the threat posed to their livelihoods.

Community discontent has risen to the point where protest marches and the sabotage of mariculture activities (poaching and damage to water canals) have become common. In response, the shrimp farmers have gone on the defensive: it is thought that farm watchmen may have been responsible for the deaths of five fishermen in 1993.[6] Of course, Honduras is not the only site of conflict resulting from shrimp farming: the *Fundación Natura* of Ecuador[7] and the US-based Mangrove Action Project recently placed a call in the Ecuadorean press and across the Internet for the boycott of 'environmentally-damaging' shrimp. This chapter will describe the trends occurring in southern Honduras and the factors which could reduce the negative effects of aquaculture. It will tell the story of a virtually powerless group of poor resource users – fishermen and gatherers – struggling for control of natural resources against the well-connected shrimp farmers arriving in the zone. Each side in the debate has organized a formal association and coalitions of international and national supporters. CODDEFFAGOLF (Committee for the Defense of the Flora and Fauna of the Gulf of Fonseca) is the grassroots environmental group incorporating a large section of the disgruntled fishermen while ANDAH (National Association of Aquaculturalists of Honduras) is a lobbying organization representing leading shrimp farmers. The outcome of the Honduran story to date has not been entirely negative. Eight years ago the fishermen faced a colossus of economic and political power; today there is at least an awareness of the environmental values of wetlands and a respect for traditional use rights in southern Honduras. To some extent, David has tamed Goliath. Whether this small measure of success can be sustained depends largely on whether shrimp projects are able to provide attractive benefits to local communities to offset the far-reaching changes which have occurred.

New exports and community conflict

Southern Honduras has undergone a number of changes as a result of of mariculture. Shrimp farms are usually built on mangrove areas or salt flats with unclear boundary lines; vegetation must be cleared to build access roads, pumping canals and grow-out ponds. Initial estimates suggest that 2,100 hectares of mangrove have been destroyed in southern Honduras by shrimp farms,[8] while other estimates place the figure at 4,300 hectares.[9] This destruction has reduced the supply of fish species and forest products for resident gatherers and fishermen. Another thorny problem is the work of seed gatherers who are indirect employees of the shrimp farms. Artisanal fishermen believe that their catches have fallen since the arrival of these larva collectors.[10] Shrimp larva gathering is a unique activity in which human effort causes more harm to the fish fry by-catch than the shrimp itself. It has been estimated that for every shrimp larva caught, five fish fry of other species are destroyed.[11] Finally, at the end of the cycle, some farms dump waste water into the common estuary from which others will be pumping water to start the process again.

But the most hotly contested of the changes brought about by mariculture has been the fencing off of communal lands for private farms. During the shellfish boom of the late eighties, there was a rush to acquire wetlands (salt and mud flats and mangroves) along the Honduran Pacific coast. To encourage investment in the industry, the Honduran government granted quasi-private property rights over state-owned coastal lands. Since 1983, the Honduran Tourism Ministry (SECTUR) and the General Direction of Fishing (DIGPESCA) have granted renewable long-term leases for over 25,000 hectares to shrimp exporting companies at the rental rate of $5 per hectare per year. Although these wetlands had been uninhabited up until that point, they had been used extensively by coastal villagers for collecting shellfish and wood products. The lack of suitable remaining land has since prompted the government to declare a moratorium on new concessions. But the effects of the leasing programme remain; the change in property rights replaced the claims of traditional users and in some instances caused evictions and hardship for coastal residents.[12] Enclosure of some Honduran 'winter lagoon' areas also has been harmful to traditional fishermen and gatherers by making them travel further afield to work.

Given the absence of agricultural insurance facilities, businessmen interested in starting shrimp farms argue that the concessions offered to the industry by the government are needed to reduce mariculture's high investment costs and the large production risks.[13] They also argue that the 'marginal', 'saline', 'unused' coastal wetlands are often considered worthless for any activity other than aquaculture.[14]

In reality, the power imbalance between the parties competing for wetlands access has been the decisive factor in the change of property rights, with artisanal fishermen coming out as losers in the process. Reported stock holders in the larger shrimp farms include the ex-President of the country, the ex-President of the Central Bank, the brother in-law of the current President, three military colonels and the director of the military pension board, as well as numerous leading bankers and industrialists. These actors have a common interest in export promotion, in contrast to artisanal fishermen who are providing for the local market. Under the current tenure arrangement, shrimp farmers are not liable for the environmental damages caused by their operations; they enjoy the rights and privileges to use water and fishery resources as desired. The fishermen, on the other hand, are obliged to respect the fences erected by the shrimp farms and would have to pay bribes or go through long bureaucratic procedures to stop the pollution and deforestation resulting from mariculture.

Yet despite the negative environmental and social side effects, the mariculture boom continues worldwide. Indeed, it is surprising that more sabotage and violence have not occurred. This is why the negative effects must be considered alongside two mitigating factors: the ability of the communities to lessen the negative effects through their self-organization and the employment generated by the shrimp industry (even if these employment prospects may be short-term).

Environmental activism in southern Honduras: David changes the rules of the game with Goliath

Historically artisanal fishermen have been very difficult to organize because of the independent nature of their operations.[15] But the recent mariculture threat has prompted a wave of initiatives. In 1988 a group of fisherfolk, university students, biologists and other professionals united to form CODDEFFAGOLF. Today it represents over 5,000 fishing families from twenty-two villages. CODDEFFAGOLF has sponsored numerous marches, road blockages and international solidarity campaigns to protest against the social and ecological impacts of shrimp farming and to push for the environmental monitoring which the government has failed to undertake. The most famous actions were the occupation of the bridge into the departmental capital of Choluteca and the land occupations to stop tractors using shrimp farm access roads.[16] The key demand of these protests is that community control of wetlands should be retained to promote a more sustainable and equitable use of resources.

CODDEFFAGOLF also produces a monthly newsletter and a weekly radio programme, and organizes talks and environmental education programmes in schools. It promotes ecotourism and liaises with national and international environmental organizations. The organization also sponsored legislation to create six winter lagoons as wildlife reserve areas. Its most controversial activities include denunciation of illegal mangrove cutting and vigilance in defence of community access rights. CODDEFFAGOLF representatives photograph and publicize environmental damage, undertake inspections of farm construction sites and artisanal fishing areas, and lobby government officials to protect community access rights.

The organization has received grants from several US and European donors (including Friends of the Earth, World Wildlife Fund, and the Inter-American Foundation) as well as two respected prizes: the 1992 Global 500 Prize given by the UN Environmental Program and the 17th Annual J. Paul Getty Prize coordinated by the World Wildlife Fund. These funds have allowed the organization to purchase its own building and training centre and embark on small-scale development projects for its members. A rotating loan fund has been started for promoting sustainable basic grains cultivation, agro-forestry nurseries and small-scale salt and shrimp farms. These projects will help communities secure land access and exploit the new mariculture technology for their own purposes.

However, worsening tidal flow patterns in the Gulf of Fonseca have made expensive pumping systems a prerequisite for any shrimp farm. Most small-scale salt-shrimp efforts supported by USAID were not successful in the early eighties,[17] and some of the CODDEFFAGOLF projects have encountered setbacks. A few project members have sold off their pond land to large-scale mariculture farms.[18]

But despite the abandonment of some of the small-scale projects and their poor image in some shrimp assocation circles, the initiation of the small farms has inherent benefits nevertheless. As a project report states, 'with the development of their productive activities the communal groups and families are reaffirming their *rights* to the land.' The very fact that those abandoning the projects received monetary compensation for their land claims (rather than having the land taken by force) shows that artisanal fishing communities are now a force to be reckoned with. CODDEFFAGOLF's intense publicity of mariculture's impacts has changed the rules of the game, and now customary use rights and land access must be recognized. In addition, CODDEFFAGOLF's activities have challenged some shrimp farmers to change their actions and improve the public image of the industry. The shrimp farmers association, ANDAH, for example, has presented an area management plan to international donors and is drawing up a 'Mariculture Business Ethics Code' to provide for 'in-house' regulation of its members with respect to deforestation and water pollution.[19] Questions over the advantages of community-based management or corporate control of wetland areas are sure to arise in upcoming fora. The actions of CODDEFFAGOLF have served to open the debate over alternative forms of property rights and natural resource tenure for sustainable environmental management.

Employment generation in mariculture: Goliath secures local allies and divides the masses

Parallel to environmental activism, employment generation in the mariculture zones is another important trend which may affect CODDEFFAGOLF's long-term success. The creation of jobs in a country plagued by high unemployment rates has been taunted as the key local benefit of mariculture and non-traditional exports in general. Shrimp farms have supposedly created 12,000 or more jobs in the southern region.[20] In the eyes of many analysts and policy makers, this employment may serve as a form of compensation to those who have lost access rights as Honduras takes the necessary steps towards modernization and 'the transition to modern capitalism'.[21] [22] Indeed, local benefits such as job creation may take the wind out of the sails of the environmental collective action.[23]

Shrimp employment opportunities for men include land clearing, larva gathering, harvesting and permanent work such as fish feeding, pond cleaning, and administration. Women work in the post-harvest tasks of deheading and packing shrimp. The industry also claims to have encouraged indirect employment in the production of ice, feeds, cartons, and heavy cargo transport. To date, about thirty per cent of the shrimp jobs are permanent, and fifty per cent of the positions are direct hire. The farms often give repeated three-month verbal contracts so that a worker remains in the 'permanent status of a temporary worker'. [24 25] Actual job creation probably is less than one *full-time equivalent* job per hectare.[26] No unions have been formed on any of the shrimp farms or within the shrimp packing plants; instead, the industry management has promoted 'solidarity associations' along the Costa Rican model. These worker clubs include discount stores, sports activities, and savings programmes for permanent employees. In return, the employees are not allowed to participate in labour unrest but are obliged to cooperate with company management.[27]

These 'permanent job with benefits' packages are dividing the labour force into privileged and non-privileged workers. Invariably, managers share out juicy contracts amongst selected workers while using the threat of unemployment as a worker-discipline device and a means to ensure company loyalty.[28] Complex permanent work contracts ultimately prevent sabotage of the production process and legitimate the skewed property rights regimes from which they arose.[29] Shrimp farming employment is best analysed using the village-level statistics I collected during a year-long period of fieldwork.[30] One of the most important findings is that some shrimp farm workers do receive salaries above the going rural wage and despite the new labour demands, there has not been a noticeable increase in the rural going wage.[31] Shrimp farms pay day labourers 17.5-21 lempiras ($ 2.5-$3), while workers on cattle ranches or corn plots receive only 10-15 lempiras ($ 1.43-$2.14) per day. And *real* agricultural wages in the zone may have fallen below their 1988 levels.[32] Unemployment remains, and shrimp farms have a large pool of potential labourers available from which to recruit.

Some of the lucky job beneficiaries are ex-fishermen or landless labourers contracted to gather the wild larva seed used to stock the shrimp farm ponds. This has been a lucrative source of employment for a very poor segment of the coastal poor. Despite the negative environmental side-effects of this work, it offers a high, relatively stable income which some want to protect. The larva gatherers logically resent environmental activists' attempts to halt shrimp farming. A parallel group of fishermen and larva gatherers actually came together briefly to make public statements against CODDEFFAGOLF. As one irate larva gatherer commented, 'we have rejected CODDEFFAGOLF's offers because they would take away the work from which our families live...' [33]

As confirmed by my visits to villages in the departments of Choluteca and Valle, the ultimate effect of mariculture job creation may be to divide coastal residents into those who favour and those who disapprove of shrimp farming, depending on their employment. In July 1991, for example, a group of fishermen captured the nets and destroyed the larva seed catch of other fishermen employed by a large mariculture firm. Such conflict among groups of the rural poor ultimately serves to allow shrimp farming to continue and environmental conflicts to persist.

In the intermediate term, the impact of the job creation will also depend on the evolution of the coastal ecosystem. Declining water quality and bankruptcies have led to several shrimp farm closures and worker lay-offs in Thailand and Taiwan.[34] Meanwhile, innovation and new technoglogies for larva seed collection may assure the industry's survival but could change the amount of jobs available. Already, hatchery-produced seed is being promoted as a substitute for wild seed with seven hatcheries due to open in southern Honduras in 1997. The number of people involved in larva gathering is therefore likely to decline.[35]

Conclusion:
Goliath stands firm, but treads more carefully

The shrimp farm expansion in southern Honduras mirrors the trends of mariculture growth worldwide. Yet in this case the unusual amount of environmental activism in the region has made the country a focus of attention. The advocacy of the non-governmental organization CODDEFFAGOLF has served to reduce the negative of effects of the new industry's expansion. A greater recognition of traditional use rights has been achieved, and some shrimp farmers have been stimulated to undertake more environmentally-sound actions towards common natural resources.

However, (following the insights of Vivian (1992)), the Honduran case shows that the sustainability of such environmental activism largely depends on the employment prospects offered by shrimp farming. Mariculture will not only survive but will appear attractive to coastal communities if shrimp export projects can create sufficient jobs to compensate a proportion of the rural population adversely affected by the enclosure of wetlands and by the environmental damage it has caused. If the industry can manage to police the negative excesses of its members, it will appear even more attractive to local communities. Such a scenario would clearly reduce CODDEFFAGOLF's support base.

On the other hand, if the industry's environmental degradation worsens or if the switch to capital-intensive technologies continues, then mariculture employment will decline. Over time, many disgruntled exlarva gatherers could become sympathetic to CODDEFFAGOLF's concerns, triggering an upswing in local environmental activism. Either way, the dynamics at play in southern Honduras hold important lessons for shrimp farming and indeed for the role of non-governmental organizations representing poor people throughout the Third World.

[1] R. Perdomo and H. Noe-Pino, 'Impacto de las exportaciones no-tradicionales sobre pequeños y medianos productores: casos del melón y del camarón', *Documento de Trabajo No. 4*. UNAH, Tegucigalpa: Postgrado Centroamericano en Economia, 1992.
[2] FPX. *Annual Report*. San Pedro Sula, Honduras: Federation of Honduran Non-Traditional Export Producers, 1991.
[3] T. Titenburg, *Environmental Economics and Policy*, Harper Collins, New York, 1994
[4] The environmental-social conflicts between artisanal fishermen and mariculture enterprises have been noted in other parts of the Third World. C. Bailey, 'The Social Consequences of Tropical Shrimp Mariculture Development', *Ocean and Shoreline Management*, 11: 31-44, 1988; L. Smith and S. Peterson, *Aquaculture Development in Less-Developed Countries: Social, Economic and Political Problems*. Boulder, Co: Westview Press, 1986. For other accounts of the Honduran case, see J. Dickenson, J. et al. 'Study of the Honduran Shrimp Industry.' (Gainesville, Fl.: Tropical Research and Development, 1988) and P. Vergne, et al. *Environmental Study of the Gulf of Fonseca*. Gainseville, FL: Tropical Research and Development, Inc., 1993.
[5] J. Vivian, 'Foundations for Sustainable Development: Participation, Empowerment and Local Resource Management', in D. Ghai, D. and J. Vivian, *Grassroots Environmental Action*. London: Routledge, 1992.
[6] *El Heraldo*, 17 Nov 1993.
[7] In Ecuador, shrimp farm construction has led to the destruction of an estimated 41,000 hectares of mangroves. See P. Parks and M. Bonifaz, 'Nonsustainable Use of Renewable Resources: Mangrove Deforestation and Mariculture in Ecuador', *Marine Resource Economics* 9: 1-18, 1994.
[8] Vergne, et al., 1993
[9] X.A. Ramírez, 'Coddeffagolf: Los defensores de los manglares del Golfo de Fonseca, Honduras.' *Revista Forestal Centroamericana* 9:27-32, 1994.
[10] Although there is unclear scientific evidence linking shrimp farms to fisheries declines, many fisher people *perceive* that the shrimp farms are responsible for the observed declining catches and earnings. See Vergne et al, 1993.
[11] G. Foer, 'Profile of the Coastal Resources of Honduras', pp.165-197 in G. Foer and S. Olsen, *Central America's Coasts: Profiles and an Agenda for Action*. Narragansett: Coastal Resources Centre, University of Rhode Island.
[12] This follows traditional enclosure movements in which individual property areas are fenced off and common use claims are ended, even though legal ownership of the soil has not changed.
[13] R. Kingsley, 'Aquaculture – Understanding the Risk Factors', *Infofish Marketing Digest*, 6: 17-18, 1986.
[14] *La Prensa*, 22 April 1991.
[15] J. Kurien, 'Ruining the Commons and Responses of Commoners: Coastal Overfishing and Fishworkers' Actions in Kerala State' in D. Ghai and J. Vivian op. cit.
[16] *La Prensa*, 26 March 1993.

[17] FPX, 1991
[18] CODDEFFAGOLF (Committee for the Defense of the Flora and Fauna of the Gulf of Fonseca). 'Progress Report to the Inter-American Foundation for Project Ho-194', Choluteca, Honduras, 1994.
[19] ANDAH (National Association of Aquaculturalists of Honduras). 'Shrimp Culture: A Positive Support to Honduras', mimeo, reprinted in *Hablemos Claro*, # 3, 7/31/90, Tegucigalpa, Honduras. 'La Industria Camaricultura Buscando Asegurar la Viabilidad a Largo Plazo', Forum Proceedings, Choluteca, Honduras, 1994.
[20] R. Chamorro, et al, 'Plan de Desarrollo de Producto Camarón 1992-1993.' (San Pedro Sula, Honduras: Federation of Non-Traditional Export Producers, 1992).
[21] John Sanbrailo interviewed by Sandy Tolan on 'All Things Considered', National Public Radio, 18 April 1992. Sandy Tolan: '[John Sanbrailo] acknowledges potential social and environmental costs and says with any change of this scale, there is bound to be some upheaval. But it's worth it, he says, for the long-term benefits that will generate income and jobs.'
[22] Indeed the academic debate over previous enclosures of the commons has centred on whether changing land rights transform agricultual production to raise labour demands and mop up surplus workers or whether the shift drives more of the rural poor to the cities. See N. Crafts, 'Enclosure and Labour Supply Revisited.' *Explorations in Economic History,* 14: 172-183, 1978. (Crafts, 1978).
[23] Vivian, 1992, op. cit. p.69.
[24] Vergne et al, op. cit. 1993.
[25] Even though shrimp farms operate year-round, a specific task position may be filled by several casual labourers during the course of a production cycle. Permanent workers are hired for the duration of a growing cycle (4-5 months) and are eligible for severance pay and other legal benefits.
[26] See D. Stanley, 'The Welfare Effects of an Export Boom: Land Enclosure and Labour Market Segmentation.' Paper presented at the Wisconsin Workshop on Sustainable Development, UW-Madison, Madison, Wisconsin, April 15-17, 1994.
[27] For analyses of the recent growth of solidarity associations in Costa Rica and Honduras, see J. Flores, *El Solidarismo desde adentro.* (San José, Costa Rica: ASEPROCA, 1989) and H. Hernández, *Solidarismo y Sindicalismo en Honduras.* (Tegucigalpa, Honduras: Federación Unitaria de Trabajadores de Honduras, 1991).
[28] G. Hart, 'Interlocking Transactions: Obstacles, Precursors or Instruments of Agrarian Capitalism', *Journal of Development Economics*, 23 (1986), p. 177-203.
Hart, 1986; C. Shapiro, and J. Stiglitz, J. 'Equilibrium Unemployment as a Worker Discipline Device', *American Economic Review*, 74(3), p. 433-444, 1984.
[29] Hart, op. cit. 1986.
[30] Repeated visit record-keeping and formal interviews of 220 households were collected over a fifteen-month period in three villages affected by shrimp farming to document the trends of land use changes, employment generation, and incomes.
[31] Stanley, op. cit. 1994.
[32] No published studies have documented agricultural wage changes in Honduras, but the author's interviews with informants in 1993 suggests that daily wages for common agricultural tasks have not kept up with inflation (currently 25% a year).
[33] *La Tribuna*, 29 July 1991.
[34] S. McClellan, 'After the Golf Rush', *Ceres* 13: 17-21, 1991.
[35] J. Espinoza, Biologist, ANDAH (National Association of Aquaculturalists of Honduras), Tegucigalpa, Honduras, Interview June 29, 1993.

Confronting Haiti's environmental crisis: a tale of two visions

Charles Arthur

On the short flight from Miami to the capital, Port-au-Prince, the view from the aeroplane window provides incontestable evidence of the magnitude of Haiti's environmental crisis. Along the western coastline, where the rivers empty into the sea, vast brown slicks discolour the blue Caribbean waters. This is Haiti's fertile topsoil, washed off the hillsides by occasional heavy rainfall, and carried away by streams and rivers. It is estimated that nearly forty million tons of soil, representing over ten thousand hectares of arable land, are lost through erosion each year.

The erosion of topsoil, the layer containing nearly all the nutrients that nourish plant life, has resulted in a severe decline in agricultural productivity over the last three decades. Soil erosion has also had knock-on effects in the form of silted up rivers that damage irrigation systems and cut output from hydroelectric plants, and the sedimentation that has decimated coastal fish stocks. A century and a half of plunder by loggers and charcoal producers has reduced forest cover in Haiti to less than three per cent of the total area. Mountainsides that were once thickly wooded are almost totally denuded of trees. Without the tree roots to retain it, the soil is highly vulnerable to erosion, and Haiti's bald mountains are now scarred by landslips and exposed rock outcrops. Adding to the problems caused by erosion, the lack of trees has led to a reduction in rainfall. Already parts of the country's north west department have degenerated into a vast dust bowl where crops will not grow. Hundreds of thousands of people must depend on daily meals provided by international relief agencies. Elsewhere, across the country, the human cost of Haiti's looming environmental catastrophe is apparent in the malnourished and impoverished peasant population. Every year tens of thousands of peasants are obliged to quit the countryside, and move to the capital and other towns in search of work.[1]

Erosion and deforestation: history

The causes of this disastrous state of affairs can be traced back to the birth of the Haitian nation. When Columbus landed in 1492 he described the island he called Hispaniola (now Haiti and the Dominican Republic) as an earthly paradise abounding in natural riches. Under the Spanish the genocide of the indigenous Arawaks and the importation of slaves from west Africa established the basis for a plantation system. However, it was not until the eighteenth century, when the western part of the island was controlled by the French, that significant agricultural development occurred. Expansive plantations growing sugar-cane were developed on the coastal plains, and coffee was planted on the lower mountain slopes. On the back of slave labour, Haiti became the most prosperous colony in the world. Yet with the colonists and their slaves concentrated near the sea ports, the mainly mountainous interior of the country remained largely unscathed.

All this was to change with the successful slave uprising that began in 1791 and culminated in the creation of the world's first black republic in 1804. The victorious generals rewarded themselves with the land formerly owned by the French colonists, but attempts to preserve the plantation system were fiercely resisted by the newly freed slaves. They associated the plantation with slavery, and wanted their own land to farm as individual plots.[2] Short of re-imposing slavery, Haiti's new leaders had little alternative but to abandon the plantation system. The land was divided and distributed to their closest supporters, who then kept, rented or sold it. Ex-slaves rented or squatted small cultivated plots, or began to clear new lands on the hills and mountainsides for their own use.

The collapse of the plantation system and its replacement by peasant smallholdings meant that the pattern of land tenure in Haiti was quite different to the *latifundio* estates and agro-export systems of the rest of Latin America. However the development of a land-owning peasantry did not in itself create the conditions for the contemporary environmental crisis. The crucial factor was the type of political and economic development that followed independence.

In the early nineteenth century the small group of relatively wealthy *mulattos* and former revolutionary officers, lacking both the capital and the labour for the development of agriculture, turned instead to commerce and the control of the state customs and tax mechanisms. In a short time an urban-based elite grew rich and powerful. Meanwhile the peasant majority farmed their smallholdings to produce mainly subsistence crops. With each new generation existing plots were sub-divided, and new areas in the mountains were cleared for cultivation.[3] Haiti developed into a starkly polarised society. The mass of the population toiled on small plots producing for themselves and the local market. In order to raise cash, peasant farmers set aside part of their plots to produce coffee, indigo and cocoa to sell to the merchants who sold on to the foreign market. The urban elite not only pocketed the profits from this export trade, but also controlled the state that relied almost totally for revenue on taxes paid by the peasantry. In this way a tiny dominant class extracted resources from the countryside while failing to provide any reinvestment or help for the peasant producers.[4]

This exploitative economic system, combined with a political structure responsive only to the interests of the elite, established the conditions in which growing numbers of peasants were obliged to move higher up the mountains in search of new land and wood for fuel. During the latter half of the nineteenth century pressure on the land, and in particular its tree cover, increased as a result of the commercial exploitation of forests, especially with the export of mahogany and logwood. The industrialization of sugar refining at the end of the century also had a negative effect. New and larger sugar-cane plantations were created, and there was a growth in the consumption of wood fuel in the refinery boilers.

The first US occupation of Haiti (1915-34) contributed significantly to the opening up of the interior. A new constitution written by Franklin Roosevelt, while secretary of state of the US navy, abolished the law forbidding foreign landholding, and over 250,000 acres were leased to North American firms.[5] The construction of roads and bridges, often built by forced labour, opened previously isolated areas of the country to domestic and foreign land speculators. Dispossessed peasants, and those wanting to avoid the US Army-run labour gangs, added to the numbers already encroaching on the mountainsides. By 1950 the population had grown to three and a half million, of which approximately eighty per cent were peasants. In the absence of any other source of fuel, trees were felled in increasing numbers, both for the peasants' own use, and to supply a growing demand for charcoal. By this time too the towns, especially Port-au-Prince, had grown in size, and consumed ever greater quantities of charcoal and wood fuel for household and light industrial use.[6] With a fast growing population of poor peasants squeezed by a predatory state interested only in extracting revenue without offering the rural sector anything in return, all the elements for severe environmental decline were in place. It took the father and son Duvalier dictatorship (1957-1986) to push Haiti to the brink.

In the process of eliminating all opposition to it, the regime of 'Papa Doc' Duvalier had a direct impact on the environment. Haiti's rapidly dwindling tree cover was further depleted by the tactic of razing many hectares of forest to deprive potential guerrilla insurgents of places to hide.[7] Less direct but more damaging still for the environmental situation was the deterioration of the agricultural sector under the Duvaliers. The corrupt and 'kleptocratic' form of government that had characterised previous administrations was taken to new extremes. Taxes, levies and duties were increased, and the Treasury became little more than a bank account for the Duvalierists' personal use. By refining and extending the extractive power of the state, the Duvaliers intensified the pressure on the peasantry and the land.

Farming methods had hardly changed over centuries, and typically the Haitian peasant's only tools remained hoes and machetes. Division of smallholdings through inheritance meant that the majority of Haiti's arable land was made up of plots of less than three acres. As a consequence of the continual need to produce crops in the simplest way on small plots, the land was consistently overworked. The peasant could not afford to leave land idle nor allow trees to remain where crops could be planted. On the steep mountain slopes such intensive production and lack of trees led to soil erosion, and declining crop yields.[8]

The American plan

By the 1980s conditions in rural Haiti had reached nightmarish proportions. The peasantry had increased in number to an estimated four and half million out of a total population of approximately six million. With an average of over 500 people per square mile, Haiti had become the second most densely populated country (after Barbados) in all of the Americas. The inability of the agricultural sector to support this population was reflected in the rural exodus to Port-au-Prince, and in the estimated 75 per cent of the peasant population living on the edge of starvation. Previously self-sufficient in terms of food production, Haiti had become increasingly dependent on food imports.[9] The issues of environmental degradation and economic decline were inextricably linked to each other, and to the wider political context. Responses to the crisis focused on the need to revive and develop the country's economy, and two divergent and conflicting strategies began to take shape.

The approach favoured by the region's dominant economic power, the United States, and promoted by its Agency for International Development (USAID), as well as by the International Monetary Fund (IMF), stressed the one comparative advantage that Haiti possessed: cheap labour. Accordingly, initiatives proposed and backed by these powerful institutions tended to focus on the promotion of Haiti's emergent urban business sector, and the development of light industrial assembly plants. Beginning in the early 1970s several hundred, mostly US-owned, companies established assembly plants in the Port-au-Prince area. With regard to the rural sector, USAID concentrated on the provision of food aid and support for the production of export crops, in particular, coffee.[10]

This strategy to re-orientate the economy away from reliance on the traditional peasant economy was severely criticized by the grassroots organizations that burst on to the Haitian political scene on the collapse of the Duvalier dictatorship in 1986. Organizations of peasants, slum dwellers, students and workers saw the US approach as part of a plan to make Haiti economically dependent on its northern neighbour.

Hostility to what was popularly referred to as the American Plan was stirred by the US-supervised Creole pig eradication programme (1981-1983). Following an outbreak of swine fever, USAID organized the slaughter of the entire Haitian pig population on the grounds that the disease could spread to the US. For the Haitian peasantry the pig was an essential economic asset, easy and cheap to keep, and an important source of cash in an emergency. Promised compensation often never materialised, and attempts at repopulation with a US pig breed, completely unsuitable to Haitian conditions, also proved a dismal failure. The eradication programme was a devastating blow to the peasant economy and to the environment, as peasant households turned to charcoal production when in need of cash. The Creole pig episode was portrayed as an example of US manoeuvres to hasten the destruction of the peasant economy with the aim of making Haiti dependent of US food imports.[11]

An alternative vision for Haitian agriculture

The alternative vision was provided by the peasant organizations that flourished in the late 1980s. Amongst others, the peasant movements of Papaye, and Milot, and the 'Heads Together' groups of Jean Rabel, represented a collective response to the problems of the rural population. They organized cooperatives, credit systems, and take-overs of idle or stolen land, and built silos to stockpile grains like rice and maize. Peasant leaders stressed the need to involve the peasants themselves in the resuscitation of the rural economy that would necessarily involve steps to restore the environment. Small groups of peasant farmers, under the umbrella of the regional peasant organizations, began to discuss and implement soil conservation techniques and to set aside land for reforestation.

A new constitution, approved in March 1987 after extensive public consultation, acknowledged both the need for agrarian reform and for moves to protect the environment. Agrarian reform, and in particular the resolution of disputes over land ownership, were seen by peasant organizations as a key to arresting further environmental decline. Insecurity resulting from the inconclusive and confusing state of land deeds and titles dating back to the post-Revolution period left peasants feeling vulnerable to eviction and therefore unwilling to plan for the future. In this context the benefits of leaving a tree sapling to mature were often outweighed by the need for quick cash that could be made by cutting it down for charcoal.

Real progress remained dependent on changes to the political system. Successive military dictatorships following the fall of Duvalier continued to block all moves to alter the status quo in favour of the peasant majority. In 1990 elections supervised by the United Nations were transformed by the presidential candidacy of the radical priest Jean-Bertrand Aristide. Hundreds of thousands of peasants registered to vote in Haiti's first ever meaningful election, and Aristide defeated the US-favoured candidate by winning 67 per cent of the vote. The democracy that Haiti's peasants hoped would permit a government with a genuine interest in resolving the rural crisis was short-lived. Aristide was overthrown by a military coup d'état. The three-year coup period was a disaster for the environment. Peasant organizations were a primary target for military repression. Their leaders and organizers were hunted down and killed, and their projects and networks destroyed. International aid organizations, many involved in reforestation projects, were forced to suspend their programmes because of the intensity of the repression.

Compounding matters were the effects of the half-hearted economic embargoes imposed by the Organization of American States and the United Nations. The embargoes failed to force the Haitian military to relinquish power while sending the living costs of the poor rocketing. Where all the trees had been felled, peasants searched for roots and even felled mature fruit trees to sell to charcoal producers. The long-term benefit of a tree as a producer of fruit and retainer of top-soil was sacrificed in the struggle just to stay alive. Butane and propane gas were exempt from the embargoes, but petrol shortages impeded the distribution of these cooking fuels, and the demand for charcoal increased accordingly.

Export agribusiness versus food self-sufficiency

With the return of democratic government following the US invasion of September 1994, the issue of economic development and the prospects for the environment were once again under scrutiny. Contrasting proposals continued to be advanced by, on the one hand, the international financial institutions, the IMF and the World Bank, together with USAID, and on the other, Haiti's peasant organizations. In August 1994, as the quid pro quo for his return to Haiti, President Aristide's representatives signed agreements with the multilateral lending institutions that committed the country to a Structural Adjustment Programme (SAP). The main points of the SAP were the privatization of the state-run industries, the lifting of import tariffs, and the promotion of the export sector. In return for carrying out this re-orientation of the economy, the Haitian government was promised US$1.2 billion in aid.

In the view of the country's peasant organizations, the Structural Adjustment Programme underpinned the realisation of the American Plan and spelt disaster for Haitian peasants. Chavannes Jean-Baptiste, spokesperson for the National Peasant Movement of the Papaye Congress

(MPNKP), Haiti's largest peasant organization,[12] claims 'the World Bank and the IMF are taking advantage of the coup and Aristide's return to implement their poison package.' He predicts the SAP will open the Haitian market to cheap foreign, especially US, food imports, thus destroying the livelihood of the small Haitian peasant. Many will be obliged to move to the cities to find work in the assembly industries, while others will have to work as agricultural labourers, producing specialist crops such as tropical fruit for export to the US market. 'We will have to eat the leftovers from their main course so we can produce their dessert.' [13]

Jean-Baptiste believes the abandonment of the rural sector to agribusiness is likely to accentuate Haiti's environmental decline. Independent experts support his view that only the peasantry has the will, the expertise, and the capacity to implement the measures necessary to halt erosion and begin effective reforestation.[14] Following President Aristide's return to Haiti, the peasant organizations began to re-build their structures, destroyed during the coup period. They re-launched programmes of reforestation (in 1995 eight million tree saplings were planted by the MPNKP alone), promoted soil conservation, set up credit schemes, and attempted to reintroduce the Creole pig. In the north of the country the MPNKP pressurized the government into giving 6,000 *carros* (1 carro = 1.29 hectares) to a peasant group which used the land in the following way: 1 carro was allocated to each peasant for their individual use; 1,000 carros were set aside for reforestation; and 2,000 were allocated for collective use, including collective animal husbandry and manioc cultivation. Some of the land will also be used for a school, a clinic, a sports field, collective mills for grinding manioc, and a sugar-cane refinery.

Jean-Baptiste contrasts the peasant organizations' approach to the environment to that of the US-sponsored non-governmental organization.

> USAID finances a lot of reforestation projects through the Pan-American Development Foundation (PADF). These projects are centralised – they send trees out from the nurseries to people around the country without doing any consciousness raising around the issue of reforestation or providing any technical training. Peasants get given a little money to plant the trees but I think that even in PADF's own evaluation eighty to ninety per cent of these trees don't survive. USAID also works through the CARE organization which has reforestation programmes but they are not part of a larger agricultural programme. They try to show their environmental concern but in reality it doesn't have any real impact that would support real development of production in the agricultural sector or development of the country in general.[15]

But the ability of the peasants' programmes to make a national impact depends on government support. 'The solution is a national economic plan,' asserts Jean-Baptiste. 'Haiti used to be almost self-sufficient but now we import 70-80 per cent of the food we consume. Two hundred thousand tonnes of rice a year are needed to feed the nation. In the Artibonite region we could produce up to 140,000 tonnes if only irrigation sustems were set up. Then there is grain. We need up to 400,000 tonnes each year. Haiti could produce this quantity within five years, and then the country could be self-sufficient.' [16]

The peasants' vision of national food self-sufficiency and a government working with the organized peasantry through the provision of credit, subsidized fertilizers, transport, lower taxes, and the distribution of state-owned lands to mixed collective and private farms, has come into direct conflict with the Structural Adjustment Programme. During the year since its restoration, Aristide's government sat uneasily between its foreign financial backers and the peasant population that brought the President to power in the first place. In February 1995 the first-ever Environment Ministry was inaugurated and announced bold plans to launch reforestation projects and campaigns of public education on the need to protect the environment.[17] The Institute of Agrarian Reform, proposed in the 1987 Constitution, finally opened in April 1995. However neither institution has made much headway, in common with the slow progress of reform that was a feature of the government as a whole during the year since its restoration.

The ability of the Haitian government to address the environmental crisis, and indeed most other pressing requirements, has been hampered by the immense scale of the tasks confronting it, the lack of resources, both human and financial, and above all the parameters outlined in the Structural Adjustment Programme. In October 1995 popular protests against privatization forced the resignation of Prime Minister Smarck Michel; a month later the US suspended US$4.5 million of aid because of the delay in implementing this aspect of the Structural Adjustment Programme. The future direction of Haiti's economic development, and prospects for the environment thus remain highly uncertain.

[1] For an overview of Haiti's environmental situation see: James Ferguson, 'Desert in the Making', *New Scientist*, 19 May 1988; Amy Wilentz, *The Rainy Season: Haiti since Duvalier*, Jonathan Cape, 1989, pp. 243-284; Federation des Amis de la Nature 'Appel aux écologistes et aux institutions engagés dans la protection de l'environnement', Port-au-Prince, 6 January 1992; Paul Wallich, 'The Wages of Haiti's Dictatorship', *Scientific American*, December 1994; Mats Lundahl, *Peasants and Poverty: a Study of Haiti*, Croom Helm, 1979.
[2] See CLR James *The Black Jacobins. Toussaint L'Ouverture and the San Domingo Revolution*, Allison & Busby, 1980.
[3] See Mats Lundahl *Peasants and Poverty: A Study of Haiti*, Croom Helm, 1979.
[4] See Michel-Rolph Trouillot *Haiti. State Against Nation: The Origins and Legacy of Duvalierism*, Monthly Review Press, 1990.
[5] Suzy Castor *L'Occupation Americaine d'Haiti*, Henri Deschamps, 1988, p.93.
[6] *Pour la Protection de l'Environnement et un Développement Alternatif*, COPHEDA, June 1993.
[7] 'Organizers work to restore Haiti's devastated environment', Reuter, 22 April 1995.
[8] Mats Lundahl 'Underdevelopment in Haiti', *Journal of Latin American Studies* no. 23, pp 411-429.
[9] Rod Prince *Haiti: Family Business,* Latin America Bureau, 1986, pp. 43-47.
[10] 'Beyond the Mountains, More Mountains. Haiti Faces the Future' in *Voices for Haiti*, EPICA 1994, pp.24-27.
[11] Michael Hooper, 'Model Underdevelopment' in *Dangerous Crossroads*, Southend Press, 1995.
[12] In the early 1970s peasants in the Central Plateau region of Haiti began organizing and working together, and the Peasant Movement of Papaye (MPP) was born. Forced to maintain a low profile for years, the MPP expanded rapidly after Duvalier's fall in 1986. Links with other peasant organizations were made, and in March 1991 the National Peasant Movement of the Papaye Congress (MPNKP) brought together representatives of over 100,000 organized peasants. Chavannes Jean-Baptiste was the founder of the MPP and is the spokesperson for the MPNKP.
[13] Author's interview with Chavannes Jean-Baptiste, 9 October 1995.
[14] For example, Paul Paryski, Chief Technical Adviser on the Environment, UNDP in Haiti – press conference – 20 October 1994, Port-au-Prince.
[15] Author's interview with Chavannes Jean-Baptiste, 9 October 1995.
[16] Ibid.
[17] 'Can political peace help Haiti's ailing environment?', Reuter,14 March 1995. Interview with Chavannes Jean-Baptiste, October 1995.

The Greening of Cuba*

Peter Rosset

The manager of the state farm at the end of the dirt road in Villa Clara province appeared the perfect technocrat. He told me how he had fought against using oxen to prepare the land on his farm. Eventually he had no choice but to give in, since petroleum, tractor parts and tyres have virtually disappeared from the Cuban countryside in the past four years. Now he grudgingly admits that oxen have their benefits. 'Before we could only fit two planting cycles into the rainy season,' he said. 'For more than a month each year, we couldn't prepare the land because the tractors got stuck in the mud. But an ox doesn't have that problem. You plough the day after it rains, or even while it's raining if you want.' As a result, he said, the farm harvests three crops a year instead of two. Although the yield per cycle is lower, the annual yield is higher. The switch to oxen wasn't easy. Though ploughing with an ox team goes back hundreds of years in Cuba, none of the farm workers or agronomists had ever farmed with oxen themselves. 'We had to hire some old campesinos from nearby villages as consultants to teach us how to hitch up an ox team and plough with them,' the state-farm manager said. 'It's amazing how much those old guys know about farming.'

The 1989 collapse of trade relations with the former socialist bloc, together with the ongoing US embargo, have meant that Cuban farmers now have to get by with a tiny fraction of the imported machinery and agrochemicals that they once depended upon. National sugar production, for example, which in the recent past averaged about eight million tons annually, has plummeted by a half in the past two years. This comes from a decrease in acreage planted because of machinery problems, and a drop in average yields from about 64,000 *arrobas* per *caballería* to around 50,000.[1] The state farms that supply the bulk of the cane to the Héctor Molina Sugar Mill, just south of Havana, experienced just such an enormous reduction in yields the last two years. Yet a number of small farmers from whom the mill also buys cane did not.[2]

Several independent farmers who use organic farming techniques have turned in yields ranging from 100,000 to 150,000 arrobas. Another peasant who practices a traditional rotation of sugar-cane with food crops and cattle pasture, though not organic, harvested about 85,000 arrobas last year, and was the largest supplier of food crops to the local marketing board as well. Farmers like these passed unnoticed when eight million tons of sugar were the norm, but they have recently captured the attention of agronomy experts who hope to disseminate their farming methods throughout Cuba. Since these peasant producers never used large quantities of the no longer available chemical inputs, they have been better able to weather the present crisis than the state sector has.

Times are changing so fast in the Cuban countryside that the only thing you can be certain of is that what is true today will no longer be true tomorrow. Since the 1960s three things characterized Cuban agricultural policy: sugar, state farms, and a fanatical love affair with the chemical and petroleum-intensive technologies of conventional modern agriculture. Yet since the 1989 collapse of imports, the Cuban government has had to give priority to food crops, and has turned the state farms over to the workers in an attempt to stimulate productivity.[3] Cuba has also embarked upon the first national transformation in history from conventional modern agriculture to large-scale organic and semi-organic farming.[4] Cuba's farmers and substantial scientific infrastructure – both physical plant and human resources – have been mobilized to substitute autochthonous technology for the foreign agricultural inputs. Cuban-made biopesticides and biofertilizers – the products of the country's cutting-edge biotechnology – are being combined with traditional peasant practices as well as ecological pest control, large-scale earthworm rearing, waste composting, and other environmentally rational practices.

This process has not stopped Cuba from falling into the worst food crisis in its history. Some estimates of the drop in average protein and calorie intake over the past three years are as high as thirty per cent, to the point where Cubans are eating better than only Haitians and Bolivians. Because of innovations in the agricultural sector, it seems that food production is going up – though not yet by enough to compensate for the drop in imports.[5]

From the Cuban revolution in 1959 through the collapse of trading relations with the socialist bloc at the end of the 1980s, Cuba's economic development was characterized by rapid modernization and a high degree of social equity and welfare. Cubans arguably had the highest standard of living in Latin America. Yet Cuba never achieved truly independent development, as the island depended upon its socialist trading partners for petroleum, industrial equipment and supplies, agricultural inputs such as fertilizer and pesticides, and even basic foodstuffs for the population. By some estimates, as much as 57 per cent of the total calories consumed by the population were imported prior to 1989.

Cuban agriculture was based on large-scale, capital-intensive monoculture, more akin in many ways to California's Imperial Valley than to the typical Latin American *minifundio* or small-scale farm. Agrochemicals and tractors replaced human labour, leading to a rural exodus, just as had occurred in the US and other countries with industrialized agricultural systems. This production model has been showing signs of strain everywhere – including Cuba – as soil erosion and pesticide resistance lead to rising costs, and stagnant or falling yields. In Cuba, more than 90 per cent of fertilizers and pesticides, as well as most of the ingredients to formulate them locally, were imported from abroad.

When trade relations with the socialist bloc collapsed, pesticide and fertilizer imports dropped by about 80 per cent, and the availability of petroleum for agriculture dropped by a half. Food imports also fell by more than a half. Suddenly, an agricultural system almost as modern and industrialized as that of California was faced with a dual challenge: the need to essentially double food production with less than half the inputs, and at the same time maintain export-crop production so as not to erode further the country's meager foreign-exchange holdings.

In some ways Cuba was uniquely prepared to face this challenge. With only two per cent of Latin America's population but eleven per cent of its scientists and a well-developed research infrastructure, the government was able to call for 'knowledge-intensive' technological innovation to substitute for the now unavailable inputs. Luckily an alternative agriculture movement had taken hold among Cuban researchers as early as 1982, and many promising research results – which had previously remained relatively unused – were available for immediate and widespread implementation.[6]

Organic farming

Most of Cuba's agricultural soils have suffered a high degree of fertility loss and depletion of organic matter due to the past intensive use of pesticides and fertilizers. To rebuild healthy soils, Cubans are now using green manure crops as part of crop rotations, composting municipal garbage and other waste products, and undertaking the industrial-scale production of high-quality humus, using earthworms as composting agents.[7] In 1992, 172 vermicompost centres produced 93,000 tons of worm humus.[8]

Waste recycling is high on the new agenda. All kinds of waste products are being converted into animal food, energy and fertilizer. These organic by-products are collected from sugar-cane processing, cattle and sheep ranches, poultry and pig farms, food and coffee processing plants, crop residues, and municipal garbage. Liquid wastes help irrigate the agricultural fields. Sugar-cane stalks are being recycled into particle board and paper, as well as into fuel for the mills' boiling pots. Integrated pig production is a good example of how complex this recycling can become. The process begins with the collection of food scraps from workplace cafeterias, restaurants and schools. These scraps are fed to the pigs as a feed supplement. Farmers may also mix in waste from slaughterhouses, which is a good protein source. Next, the liquid and solid waste from the pigs is recycled to be used in vermiculture, in biogas generation, and even as a feed supplement for the same pigs. The stated goal is to reach zero non-recycled waste.

Cuba has a unique system of pest management in sweet potatoes – a staple of the local diet. Predatory ants are mass-reared in banana stems and introduced into the fields at the point when tuber formation begins. The ants then build their nests around the sweet potatoes in the soil, protecting them from the ravages of the sweet-potato weevil. A similar method is used in plantain plantations. Cuba now has fourteen centres for ant production scattered around the country.[9] Other centres mass-rear other insects that prey upon or parasitize various species of crop pests.

Empirical evidence from the US and elsewhere demonstrates that organic-farming methods can take between three and seven years from the initiation of the conversion process to achieve the levels of productivity that prevailed beforehand.[10] That is because it takes time to restore lost soil fertility and to re-establish natural controls of insect and disease populations. Yet Cuba does not have three to seven years; its population must be fed in the short term. Because of the urgency of the crisis, Cuban scientists and planners are bringing sophisticated biotechnology to bear on the development of new organic farming practices.

Biotechnology

In the US we are unused to hearing the words 'biotechnology' and 'organic farming' in the same sentence. We tend to think of all bio-technology in terms of releasing genetically engineered organisms into the environment, a process which poses ecological and public-health risks that are not consistent with the goals of organic farming. What the Cubans are doing is different. They are collecting locally occurring strains of micro-organisms that perform useful functions in natural ecosystems. These range from disease microbes that are specific to certain crop pests, and thus non-toxic to other forms of life, to micro-organisms that convert atmospheric nitrogen into a form that crop plants can use. These micro-organisms are then massively reproduced in order to be used as biopesticides and biofertilizers in agro-ecosystems.[11] Some of these products are available commercially in the US as well, but Cuba is way ahead in terms of the diversity of such biological preparations that are in widespread use.

Located on agricultural cooperatives, 222 artisanal biotechnology centres produce these biotech products for local use. These products are typically made by people in their twenties, who were born on the cooperative and who have some university-level training. In a sense, Cuba is demystifying biotechnology for developing countries by showing that biotechnology does not have to rely on multi-million-dollar infrastructure and super-specialized scientists. Rather, the sons and daughters of campesinos can make and use biotechnology products in remote rural areas. Industrial production of these biopesticides will soon be under way for use in larger-scale farming operations that produce for export. The labour-saving methods of biotechnology are particularly appropriate for Cuba because like the US, Cuba faces labour shortages in agriculture. Eighty per cent of the Cuban population lives in urban areas and only twenty per cent lives in the countryside, while in other countries with widespread alternative agriculture such as China, this ratio is reversed. At the same time, the Cuban government is going to great lengths – such as constructing high-quality housing and entertainment centres – to make relocating in the countryside an attractive option for city dwellers.[12]

Linking people with the land

Cuba is also radically reorganizing its production in order to create the small-scale management units that are essential for effective organic farming. This reorganization has centred on the privatization and cooperativization of the state sector.[13] Under conventional systems, a single technician can manage several thousand acres on a 'recipe' basis by simply writing out instructions for a particular fertilizer formula or pesticide to be applied with machinery on the entire area. Not so for organic farming. Whoever manages the farm must be intimately familiar with the ecological heterogeneity of each individual patch of soil. The farmer must know, for example, where organic matter needs to be added, and where pest refuges and entry points are.

In Cuba this scaling back of production units has coincided with the issue of production incentives. Several years ago planners became aware that the organization of work on state farms was profoundly alienating in terms of the relationship between the agricultural worker and the land. Large farms of thousands of acres had their work forces organized into teams which would prepare the soil in one area, move on to plant another, weed still another, and later harvest an altogether different area. Almost never would the same person both plant and harvest the same field. Thus no one ever had to confront the consequences of doing something badly or, conversely, enjoyed the fruits of his or her own labour. In an effort to recreate a more intimate relationship between farmers and the land, and to tie financial incentives to productivity, Cubans began several years ago to experiment with a programme called '*Vinculando el hombre con la tierra*,' or linking people with the land. This system made small work teams directly responsible for all aspects of production in a given parcel of land, allowing remuneration to be directly linked to productivity. The new system was tried on a number of state farms, and rapidly led to enormous increases in production.

The process of linking people with the land culminated in September 1993, when the Cuban government issued a decree terminating the existence of state farms, and turning them into Basic Units of Cooperative Production (UBPCs), a form of worker-owned enterprise or cooperative. The 80 per cent of all farmland that was once held by the state, including sugar-cane plantations, has now essentially been privatized into the hands of the workers. The UBPCs allow collectives of workers to lease state farmlands at low rent, in permanent usufruct. Property rights remain in the hands of the state, and the UBPCs must still meet production quotas for their key crops, but the collectives are owners of what they produce. Perhaps most importantly, what they produce in excess of their quotas can now be freely sold on the newly reopened farmers' markets.[14] Members elect management teams that determine the division of jobs, what crops will be planted on which parcels, and how much credit will be taken out to pay for the purchase of inputs.

The pace of consolidation of the UBPCs has varied greatly in their first year of life. Today one can find a range from those where the only change is that the old manager is now an employee of the workers, to those that truly function as collectives, to some in which the workers are parcelling the farms into small plots worked by groups of friends or *socios*. It is still too early to tell toward what final variety of structures the UBPCs will evolve.

Even before the present crisis began, Cuba had been trying to move closer towards self-sufficiency in food crops.[15] Under the National Food Program, which began in the mid-1980s, sugar estates have been required to plant food crops and raise livestock in uncultivated areas. The goal was for each farm to supply the food needs of its workers and their families. The cultivation of beans, plantains, and root crops has increased as a result, although exact figures are hard to come by.

Dynamic debate

This process of downsizing and conversion to organic farming is, of course, not taking place without controversy and setbacks. A dynamic debate is underway inside Cuba which cuts across the agricultural sector, from government ministries, universities and research centres, to farmers and associations of producers. One side argues that what is taking place should be seen not precisely as a process of conversion, but rather as a temporary substitution during a period of crisis. This viewpoint holds that once trade conditions change, agrochemical inputs should once again be vigorously used. The opposite point of view, put forth by the Cuban Association for Organic Farming among others, holds that the previous model was too import-dependent and environmentally damaging to be sustainable. People in this camp argue that the present change was long overdue, and that further transformations are needed to develop truly rational production systems.[16]

The Organic Farming Association is a non-governmental organization (NGO), a rare phenomenon in Cuba.[17] The association is playing a pivotal role in what might be called the institutionalization of the alternative model. Members are ecological agriculture activists ranging from university professors and students, to mid-level government functionaries, farmers and farm managers. They are struggling on a shoe-string budget to carry out an educational campaign on the virtues and indeed the necessity of maintaining and reinforcing the alternative model.

Opponents of such institutionalization point to the recent collapse of Cuban attempts at massive implementation of so-called Voisin Pasture Management as evidence of the inadequacies of organic farming technologies. The Voisin system was supposed to maintain dairy productivity without the widespread use of chemical fertilizers on pastures. The basic principle is as old as animal husbandry itself – the rotation of paddocks in such a way that manure is supplied to growing grasses at the precise moment when it is most needed. The Voisin system failed in Cuba, however, because it required portable electric fencing that was both in short supply and susceptible to the ubiquitous power cuts, and because the density of cattle per acre was too high. Advocates of an alternative model for agriculture point out that it was not the principle of rotational grazing that failed, but rather the way in which it was applied.

Such debate aside, what may be most remarkable about the recent changes in Cuban agriculture is the rediscovery of the traditional values and knowledge of farmers. The Ministry of Agriculture has launched a national programme to recover traditional farming knowledge, recognizing that peasants have always practiced low-input, agro-ecologically sound agriculture. Mobile seminars and workshops are taking place around the country, where farmers can meet to trade their farming secrets and to share them with researchers and government officials.

If a silver lining to the current crisis exists, it is surely the new integration of socialist values with environmental consciousness and greater individual responsibility. Roberto García Trujillo, an assistant dean at the Agricultural University of Havana (ISCAH), is the founder of the Organic Farming Association. On the weekends, he practices what he preaches, working his tiny organic farm on a patch of land inherited by his wife. While he and his son turned the compost pile one Sunday morning, he mused:

> Many people think that farming is a simple and mundane act, but they are wrong. It is the soul of any great culture, because it requires not only a great deal of accumulated knowledge, but also putting this knowledge to use every single day. Knowledge of the weather, the soil, plants, animals, the cycles of nature; all of this is used everyday by a farmer to make the decisions that have to be made in order to produce the food that we eat. To us it may seem like food comes from a factory, but in reality it comes from a culture that, generation after generation, has been created to produce that food.[18]

Conclusion

It is clearly too soon to tell if the transformation of Cuban agriculture will be permanent, or even if it will help Cuba survive its present crisis. Nevertheless, Cuba may turn out to be a model for the rest of us. Whether we live in Latin America, the US, Asia, Africa or Europe, we are all facing the declining productivity of modern conventional agriculture. As soils are progressively eroded, compacted by heavy machinery, salinized by excessive irrigation, and sterilized with chemicals, and as pests become ever more resistant to pesticides, crop yields are in decline. Meanwhile, aquifers and estuaries are being contaminated with agrochemical run-off. Organic farming and other alternative technologies are intensively studied in laboratories and experimental plots worldwide, but examples of implementation by farmers remain scattered and isolated. Cuba offers us the first large-scale test of these alternatives. Before we are all forced to make this transformation, this island nation offers us perhaps our only chance to see what works and what doesn't, what the problems are and which solutions will emerge.[19] Cuba is also carving out a path back from the de-skilled work process of large-scale industrial farming, toward a more human endeavor, engaged equally with traditional knowledge and modern ecological science.

*First appeared in *NACLA Report on the Americas*, New York, Vol XXVIII No3, Nov/Dec 1994
Tracy Lynn Ackerly assisted with this article.

[1] Cuban measurements for weight and area. 1 arroba = 25 pounds and 1 caballería = 33 acres.
[2] Until late 1993, 80 per cent of arable land was in State Farms, 11 per cent in peasant cooperatives, and 9% in the hands of small independent farmers.
[3] In 1989, Cuba devoted three times as much land to sugar as to food crops, or about 53 per cent of non-pasture arable land.
[4] For a fuller discussion of the issues (as well as data sources) presented in this article, see Peter Rosset and Medea Benjamin, *The Greening of the Revolution: Cuba's Experiment with Organic Agriculture*, Ocean Press and Global Exchange, available for $11.95 plus $2.50 postage and handling from Food First by calling 1-800-888-3314. Also see Peter Rosset and Shea Cunningham, 'The Greening of Cuba: Organic Farming Offers Hope in the Midst of Crisis', Food First Action Alert, Spring, 1994.
[5] Reliable figures are not available for two reasons. First, the compilation and release of national statistics have been curtailed during the economic crisis, and second, official figures record only that food sold through government channels, ignoring the burgeoning black market.
[6] Richard Levins, 'The Ecological Transformation of Cuba,' *Agriculture & Human Values*, Vol. 10, No. 3, 1993, pp. 3-8.
[7] Green manures are legume plant species that can be planted as cover crops to supply nitrogen to the soil.
[8] See Paul Gersper, Carmen Rodríguez-Barbosa and Laura Orlando, 'Soil Conservation in Cuba: A Key to the New Model for Agriculture', *Agriculture & Human Values* Vol. 10, No. 3 1993, pp. 16-23.
[9] Nicolas Lampkin, *Organic Farming*, Farming Press, UK, 1990.
[10] Beatriz Díaz and Marta R. Muñoz, 'Biotecnología agrícola y medio ambiente en el período especial cubano', paper presented at the XVIII Meeting of the Latin American Studies Association, Atlanta, Georgia, 10-12 March, 1994.
[11] Jeff Dlott, Ivette Perfecto, Peter Rosset, Larry Burkham, Julio Monterrey, and John Vandermeer, 'Management of Insect Pests and Weeds', *Agriculture & Human Values* Vol. 10, No. 3 1993, pp. 3-9.
[12] Peter Rosset and Medea Benjamin, 'Two Steps Back, One Step Forward: Cuba's National Policy for Alternative Agriculture', International Institute for Environment and Development Gatekeeper Series No. 46, London, 1994.
[13] For a discussion of the conversion of state farms into UBPCs, see Carmen Diana Deere, Niurka Pérez and Ernel González, 'The View from Below: Cuban Agriculture in the Special Period of Peacetime', paper presented at the XVIII Meeting of the Latin American Studies Association, Atlanta, Georgia, 10-12 March, 1994.
[14] Cuba experimented with these markets in the early 1980s, but they were shut down because of concerns about the creation of a class of middlemen. See Joseph Collins and Michael Scott, *No Free Lunch: Food & Revolution in Cuba Today,* Food First Books, San Francisco, 1984. They were just reopened in September of this year in an attempt to stimulate production through price incentives. See 'Cuba Will Allow Farmers to Sell on Open Market,' *The New York Times*, September 18, 1994.
[15] Laura J. Enríquez, *The Question of Food Security in Cuban Socialism*, University of

California Press, Berkeley, 1994; Carmen Diana Deere, 'Socialism on One Island: Cuba's National Food Programme and its Prospects for Food Security', Institute for Social Studies, Working Paper Series No. 124, The Hague, 1992.

[16] Roberto García Trujillo, *La conversión hacia una agricultura orgánica*', Asociación Cubana de Agricultura Orgánica, Havana, 1993.

[17] To contact the association, please write to: Asociación Cubana de Agricultura Orgánica, Tulipán No. 1011, entre Loma y 47, Nuevo Vedado, CP 10600, Havana, Cuba.

[18] Interview in 'The Greening of Cuba', a documentary being produced by the Institute for Food & Development Policy.

[19] See John Vandermeer, Judith Carney, Paul Gersper, Ivette Perfecto, and Peter Rosset, 'Cuba and the Dilemma of Modern Agriculture', *Agriculture & Human Values*, Vol. 10, No. 3, 1993, pp. 3-8; Miguel A. Altieri, 'The Implications of Cuba's Agricultural Conversion for the General Latin American Agro-ecological Movement', *Agriculture & Human Values*, Vol. 10, No. 3, 1993, pp. 91-92.

Puerto Rico's energy fix

Marianne Meyn

To Julia Mignucci, a woman committed to a life with dignity for all human beings.

Resistance to coal-fired power

Saturday, 1 August, 1992. The day of a historic march and demonstration by over 10,000 residents from the coastal town of Mayagüez in the west of Puerto Rico. Their fierce opposition to the building of a coal-fired power plant had brought them out in force. Proposed by the Cogentrix-ENDESA consortium, this power plant was supposed to 'get Puerto Rico out of its dark future in a clean way' but such assertions had not fooled local residents. Even the torrential rain did not frighten off the marchers; on the contrary, the crowd seemed even more united by the weather. The more creative ones chanted: 'It's better to soak than have Cogentrix' coke'. The movement against the Cogentrix plant had been initiated originally by the ecumenical ecology project, Industrial Mission of Puerto Rico.[1] It began with a local barbecue which we organized in order to talk to people about the dangers of coal as an energy source. It culminated in the biggest demonstration ever seen in this quiet town of 100,000 inhabitants.

No political or economic power could impose this polluting project on the people of Mayagüez. The project had been more heavily promoted by private business and government agencies than any other in Puerto Rican history. And yet the $15 million Cogentrix-ENDESA invested in bribes and publicity campaigns was money down the drain, thanks to the combined efforts of Mayagüez's residents. Everyone's talents were mobilized: activists gave out thousands of leaflets; artists daubed the town in murals and other visual symbols; those adept at public speaking reported on radio, television and in the press; scientists and technicians analysed the thousands of written pages on the environmental, economic and social impact of the plan in a sixteen-day public hearing – the longest in the history of Puerto Rico's environmental struggle. Priests said Sunday mass on the land proposed for the plant in the Punta Algarrobo district, next to the bay of Mayagüez; on the same land children planted trees, fisherfolk held fishing competitions and the residents entertained and educated people in environmental festivals. Those with a taste for detective work uncovered dishonest advertising by the consortium; lawyers successfully defended harassed activists such as the geology professor who was sacked for pointing out that the plant was going to be built on a geological fault;

opportunists with some conscience denounced the bribes they had been offered. Business and unions made donations to all kinds of activities; all Cogentrix-ENDESA publicity events were picketed and at Christmas time a *trulla navideña* [2] (group of carol singers) was brought to Cogentrix-ENDESA's information desk in a Mayaguez shopping centre to sing environmental songs. Protesters interrupted the Municipal Assembly to demand a hearing; they symbolically invaded the Environmental Quality Board to ask for the resignation of the president; they boycotted the public hearings of the US Environmental Protection Agency; during election campaigns they demanded that politicians take a position on the issue. Even the drunks helped out by drinking the thousands of dollars worth of beer donated by Cogentrix-ENDESA, hanging around in bars until the early hours of the morning to bad-mouth the company.

In June 1993, after nearly four years of struggle, the organization leading the movement, the People of Mayagüez for Health and the Environment, was victorious: the Cogentrix-ENDESA consortium had to surrender and withdraw the proposal. None of its manoeuvres or traps had been able to defeat the community; not the buying of the heads of agencies; not the threats about the collapse of the industrial development model; not the donations to schools, civic organizations, sports clubs, and others; not the radio and TV programmes, nor the deceitful adverts, nor drinking a glass of water with pieces of coal in it in front of the TV cameras; not injunctions to restrict the public expression of people's feelings, could break the will of the people. In spite of the fact that all the government agencies had sold themselves to the company, they had to give in to the political strength of the Mayagüezan popular movement.

To 'commemorate the growth of awareness of power in a people united in their struggle to conserve the purity of the environment,' the Mayagüezans built an obelisk on the land intended for the power plant. Two months later, the Economic Development Administration demolished the Monument of Punta Algarrobo under cover of darkness to bury this symbol of popular strength. After the defeat, the government could not stomach the sight of the Mayagüezans nor their obelisk.

Why was there such massive community opposition to a coal-fired power plant? It is obvious that those living nearby were worried about the high degree of air pollution (dust particles, nitrogen, sulphur, dioxins, radioactivity), the resulting health problems and the effects on flora and fauna, on water and soil, the disposal of toxic ashes, the discharge of pollutants into the sea, the storage of coal in the open air and other problems which would appear only half a mile away from residential areas and a school. Moreover, they were aware of the experiences of neighbouring communities like Cataño on the north coast where the Electrical

Energy Authority has been operating a gigantic thermoelectric power plant. The plant has been among the dirtiest in the US and its territories, with more than 8,000 violations of environmental laws and regulations officially recognised in 1993. For two decades local residents had been saying 'Cataño is hell'.[3] The President of the Environmental Quality Board affirmed during the same period that the only solution to the problem of pollution in Cataño would be to 'move the people out of there'.[4]

Apart from their fears about pollution, the people of Mayagüez were aware that there was no need for another power plant in Puerto Rico. In fact, the struggle against the Cogentrix-ENDESA coal-fired plant was one of many struggles by local communities against new power plants which began to break out in the early 1970s and are continuing today with the proposed coal-fired plant for Guayama. At that time it was already becoming clear that the government policy of servitude to multinational corporations was destined to fail and that local communities were not willing to put up with the effects on health and the environment of an unnecessary energy plant. The people refused to swallow the myth that energy capacity was automatically synonymous with economic prosperity.

Operation Bootstrap

The Puerto Rican government is still trying to make their myth a reality with its continued pursuit of a development policy initiated in the 1940s. This policy involves providing foreign companies with infrastructure (energy, water, land, communications, buildings and low-cost labour), tax exemptions and access to the US market in exchange for industrial and tourist trade growth which, in turn, is supposed to generate jobs. But the government has failed to recognise the health problems resulting from this strategy, let alone the high level of environmental pollution, degradation and destruction of subtropical ecological systems, the social problems resulting from a sudden transformation from an agrarian society into an industrialized one, or the vicissitudes of a dependent economy. The government has chosen to hide the fact that today unemployment in the country is as high as it was in the 1940s. That is, before the 'industrialization by invitation' programme – otherwise known as Operation Bootstrap – had even been conjured up.[5]

Operation Bootstrap pioneered one of the most rapid industrial revolutions in the world.[6] It put Puerto Rico up for sale as an industrial location for transnational corporations, principally North American. In the space of only three decades, Puerto Rico was to be transformed from an agrarian society into an industrial, urbanised one. By 1978 more than 2,000 factories promoted by the Economic Development Administration[7] were in operation, among them more than eighty pharmaceutical factories, around a hundred in the petroleum and related products sector, plus factories producing electronic components, electrical goods, and medical equipment.[8] At that time environmental pollution was already 'our daily bread', in the words of one community leader.[9] A NACLA report in 1981 synthesised the situation with an ironic adaptation of the slogan of the Popular Democratic Party which was promoting the policy of 'industrialization by invitation': Bread (foreign), Land (wasted), Freedom (denied).[10]

Drive for increased energy capacity

In the first phase of Bootstrap, the government did not need much energy capacity to serve the foreign industries. Its main concern was to invite light industries whose energy needs could be met by existing hydroelectric systems and other small electricity generators. However, by the second phase of Operation Bootstrap, energy capacity was considered to be a crucial factor. When the manufacturing industry started to stagnate in the late 1950s, the Administration of Economic Development began to throw in its luck with the petroleum industry. Despite the fact that the oil companies only employ small workforces, it was hoped the petroleum industry would create centres for industrial complexes where so-called satellite industries would generate jobs for the people. By 1955 the Economic Development Administration had itself encouraged the establishment of a gigantic state refinery, the Commonwealth Oil Refining Corporation (CORCO), which was projected in the press as a symbol of progress, 'the incarnation of Puerto Rico itself'.[11] The consumption of energy by the petrochemical factories was colossal. Just three plants, one PPG and two Union Carbide, consumed a third of the island's energy capacity, exceeding the consumption of fourteen Caribbean and Latin American countries put together.[12] Total industrial consumption went up to 74 per cent of the energy generated in the country.

The clamour for energy capacity was epitomised by the government's willingness to allow Puerto Rico to be used by the US Atomic Energy Commission for the experimental nuclear plant, BONUS (Boiling Nuclear Super Reactor), set up in the picturesque town of Rincón on the west coast in 1961. For Puerto Ricans, the Westinghouse BONUS reactor signifies one of the most tragic chapters in the history of their environmental pollution.[13] Innumerable accidents and radiation leaks occurred in this plant which had been intended as a nuclear technology exhibition centre for the whole of Latin America. A worker at the plant declared four years after its emergency closure in 1968 that 'the alarm bells used to go off continuously'.[14] Ignoring such reports, the Secretary for Health insisted that statistics showing an increase in cancer cases in Rincón were linked to people's smoking and drinking habits.[15]

The radiation leaks provoked a series of community struggles which blocked the construction of three additional thermonuclear plants proposed by the Electric Power Authority (quasi-public Puerto Rican power company)[16] and Westinghouse. These nuclear plants were to have been at the core of an expansion in petrochemical factories, satellite industrial complexes and thousands and thousands of jobs. The first River Sources Authority and Economic Development Administration energy megaproject, proposed under the name NUPLEX in 1970, seemed to be the product of an obsession with grandeur. Twin nuclear plants and two thermoelectrical ones in the Bay of Jobos (Guayama) were to generate more than 2,000 Megawatts while the heavy industry and satellite businesses which would sprout up around the plants would generate 13,500 jobs. By 1970 the Electric Power Authority had already obtained a permit from the Planning Board to build the plants and had bought the necessary equipment from Westinghouse. The Authority was trying to obtain the funds for a downpayment so that Westinghouse would start supplying the uranium (which was going to be stored who knows where).

Resistance to nuclear power

In spite of the advanced nature of these nuclear power plants, the local community and environmental groups, among them the Industrial Mission, Citizens for the Conservation of Natural Resources, the Sierra Club of Puerto Rico and the Environmental Institute, succeeded in paralysing the project in August 1972. The blocking of NUPLEX was the first big environmental blow suffered by the development politicians. The sting of defeat prompted the then Secretary of Public Works to describe the environmentalists as Luddites, 'extremist groups in the industrial revolution of eighteenth century England who destroyed machines and posed an obstacle to economic progress'.[17]

Nevertheless, the government did not lose its faith in the apparent energy miracle which the nuclear plants represented. In its passion to turn Puerto Rico into 'the junior partner of the corporate structure of the US',[18] it lined up pledges to construct seventeen nuclear plants between 1980 and 2000 on an island of 3,421 square miles.[19] Given that the US Atomic Energy Commission had finally denied permission to build nuclear plants in the Bay of Jobos because of its proximity to a geological fault rather than making a decision in principle, the Electric Power Authority saw a possible way out and tried to identify alternative sites to locate a reactor.

But local communities were having none of it. There was immediate opposition and community action when the area of Tortoise Lake (Manatí) was chosen for a nuclear plant and there were floods of protests when they cheated and expropriated 100 familes in the Islote district of Arecibo to build the North Coast Nuclear Power Plant (NORCO-NP). The opposition to this latter plant assumed national proportions and succeeded in forcing the River Sources Authority to withdraw the proposal in 1975. The equipment for the plant, already bought from Westinghouse, is still stored in the shadow of two gigantic thermoelectrical plants built illegally next to the Bay of Jobos after the NUPLEX nuclear project was not approved. In vain, the following dismal offer was circulated: 'For Sale: one 600 Mw nuclear power plant. $75.2 million. Serious enquiries only. Owner must sell before it deteriorates into scrap iron.'[20] In 1980 the government gave in to the people's demands and established a public policy which rejected the construction of nuclear plants in Puerto Rico.

Pharmaceuticals: rescue remedy?

The Economic Development Administration's policy of converting Puerto Rico into a staging post for petroleum refining was soon frustrated. This was not because of a lack of energy capacity in the island which was more than enough, according to data collected by the Industrial Mission.[21] The exodus of many petroleum factories halfway through the 1970s was a consequence of the oil crisis, signalling the end of the cheap petrol era and the advent of US tariffs on foreign petroleum. On the south coast, the petroleum factories left an enormous expanse of contaminated land, sick residents, exhausted agriculture and high unemployment. Today the installations are a rusting, industrial ghost town, symbolising the myopia with which the development model was planned.[22]

Desperate to rescue its development model, the Puerto Rican government then directed its servility towards the chemical-pharmaceutical transnationals. Of course many of these companies had already invited themselves in, lured by crystalline water from abundant aquifers on the north coast, low cost energy and a licence to discharge any kind of toxic waste into the rivers, sea, subsoil and atmosphere. To encourage the installation of yet more pharmaceutical factories and other high-tech industries, the government rehashed its attractive policy of tax breaks and offered sympathetic environmental let-outs. New factories duly arrived and piled up along the north coast, eventually expanding into other parts of the island. All the giants were represented: Schering, American Cynamid, Abbot, Pfizer, La Roche, Merck, Upjohn, Squibb, Baxter, DuPont, Monsanto and others with one or more plants.

The pharmaceutical companies, just like other high-tech industries, provide few jobs. While they appear clean and antiseptic from the outside, they have unleashed a nightmare for innumerable communities. In Humacao, the former spokeswoman of Ciudad Cristiana, a community of 450 families polluted with mercury, said shortly before the inhabitants' protests achieved their evacuation and relocation: 'many of us are living like vegetables.' Accidents, gas escapes, explosions and other incidents have kept these communities in a state of terror. Nobody knew where it would end. In Yabucoa community leaders claimed there were not enough priests to bury the dead. In Junquitos chloroform emissions from the Squibb plant caused the inhabitants to sleep in until midday. However, not all multinationals managed to establish themselves in Puerto Rico. From the late 1970s, the Industrial Mission organized more and more communities to fight against new installations by polluting transnationals.

Enter the coal-fired power plants

Although the high-tech industries consumed less energy than the petrochemical ones, and reserves were over 150 per cent, the Electric Power Authority was determined to add 900 Mw to the existing capacity of 4,200 Mw through coal-fired plants. The first project proposed was to be built next to the sealed nuclear plant BONUS in Rincón, but a wave of protests by the local community and by the pioneers of the Industrial Mission's energy struggles succeeded in protecting this small fishing town, its white beaches and its vast mangrove swamps.

In 1981 the proposal was moved further north to Aguada, but again the local communities disapproved. They insisted that there was no need for more energy and that they did not want to breathe in dirty emissions from the burning coal. They also feared the plant would tempt more factories into the area (they were already protesting against Dow Chemical). The environmentalists' educational and organizational campaign reached its peak with a march and demonstration by 5,000 activists, the largest of the environmental protests of the 1980s. The Electric Power Authority felt so threatened by this demonstration that they turned off the electricity in the whole municipality during the event. But protestors simply lit candles and ran the microphones off an emergency generator.

This broad community struggle silenced the authorities for eight years until 1989 when they invited the Cogentrix company to set up a coal-fired plant in the town of Mayagüez. This was in spite of the fact that there was still an excess of 50 per cent in energy capacity. As a result of the Caribbean Basin Initiative (CBI) and the promotion of plans for a North America Free Trade Agreement (NAFTA), many industries had moved their production operations to the Caribbean and Central America, leaving only administrative offices in Puerto Rico.

Triumph of local community action

Such a massive, militant and profoundly democratic popular movement in Mayagüez had not emerged spontaneously; it was the result of more than two decades of community organization and education encouraged by the Industrial Mission throughout the island. The movement had sought to secure local autonomy, encouraging communities to recover political spheres which historically they have delegated to governments, political parties and other institutions. Today these communities are keen to participate directly in the search for a development strategy in harmony with nature. Through their own independent organizations, they have prevented an endless number of industrial plants, tourist mega-complexes and infrastructure projects of danger to the environment.[23] Taken together, the grassroots organizations established to defend the environment represent the strongest, broadest and most consistent popular movement in Puerto Rico. It should be noted that most of the environmental actions have been headed by women, especially in the urban areas.

The action of the people of Mayagüez has added a new ingredient to Puerto Rican community action: the search for alternatives. Finding an alternative to the burning of fossil fuels was fundamental to the opposition to the coal-fired plant. After the victory against Cogentrix-ENDESA, twelve community organizations – most of them with experience of struggles related to energy projects – formed the United Environmental Front (FUA) with the objective of formulating an alternative public energy policy and fighting for its implementation. They drew up concrete proposals for conserving energy, increasing the efficiency of existing plants, co-generating, storing the excess and above all, using renewable resources such as sun, water, and wind. Since then the FUA has forced the government to declare the use of renewable energy sources a priority and to revive the Administration of Energy Affairs as an independent body for determining the country's energy policy in consultation with local communities.

Today, the movement which began with a garden barbecue is making policy.

Translated by Catherine Matheson

Ecological profile of Puerto Rico

Generation of toxic and dangerous waste (official): 1.4 million tons per annum; 400 tons per sq mile (US 73 tons per sq mile); 40 per cent exported to US; 40 per cent disposed in Puerto Rico; 20 per cent illegally disposed (1995)

Generation of toxic and dangerous waste (conservative estimate): 2.3 million tons per annum (1995)

Generation of solid waste: 2,874,000 tons per annum reach sewers (1990-91); 6-700,000 tons per annum disposed of clandestinely; the highest in Latin America.

Number of places highly contaminated by industrial waste (official estimate): 300; sites proposed for inclusion in the list of places requiring decontamination (superfund sites): 150 (1993).

Chronic cases of respiratory illnesses (official): 564,397 (16.1%) (1993).

Deaths from respiratory problems: 10 per cent (1993).

Emission of toxic substances: 0.2 tons per person, a level higher than that in the US (1988).

Cars: 1,600,000 (1 for every 2 inhabitants)

Density of roads: 13,690 miles.

Forested: 924.2 sq miles (27%) (1994).

Protected forest: 3.2% of total area.

US military bases: 90 sq miles (3%).

Contaminated beaches declared unfit for swimming (official): 45% (1994).

Polluted rivers (official): 86% (1993)

Lakes which do not fulfil water quality standards (official): 79% (1993).

Estuaries which do not fulfil water quality standards (official): 66% (1993)

[1] The NGO, Misión Industrial de Puerto Rico (Industrial Mission of Puerto Rico) was founded in 1969 by the Episcopal Church as an organizational response to the socio-economic impact of the dependent economic development model implemented in the country. Since then, it has accompanied local communities' collective actions to protect the environment and to counteract the strategies of state indsutrial growth and big private interests. (see Wilgfred Montañez and Marianne Meyn, 'Modelos de Desarrollo Capitalistas y destrucción ambiental', in Ingemar Hedstom, *La situación ambiental en Centroamérica y el Caribe*, DEI, San José, Costa Rica, 1989, pp.129 onwards).
[2] A surprise musical ambush, a typical Christmas tradition in Puerto Rico.
[3] *The San Juan Star*, 13 May 1973, pp.1 & 20.
[4] *El Mundo*, 9 Jan 1975, p.1.
[5] See Emilio Pantojas-García, *Development Strategies as Ideology: Puerto Rico's Export-led Industrialization Experience*, Lynne Rienner, Boulder, Colorado, 1990, p.51 onwards.
[6] James L. Dietz, *Historia Económica de Puerto Rico*, Río Piedras, Ediciones Huracán, 1989, p.277.
[7] Ibid., p.286.
[8] Ibid., p.275.
[9] Cielo Martín, community leader from Mayagüez.
[10] Américo Badillo-Veiga, 'Bread (foreign), Land (wasted), Liberty (denied)' in *Report on the Americas*, NACLA, vol XV, no.2, March/April 1981 p.23.
[11] Juan Manuel García Pasalqua, 'The Corco-WRA Scandal' in the *San Juan Star*, 5 Nov 1974.
[12] Trinidad and Tobago, Dominican Republic, Costa Rica, El Salvador, Guatemala, Nicaragua, Paraguay, Panama, Honduras, Bahamas, Guyana, Haiti, Ecuador and Bolivia. See Tomás Morales Cardona, *Problemas de Toxicología en Puerto Rico*, Mimeograph, 1983, p.19.
[13] National Ecumenical Movement (PRISA), *No hay mal que dure...:Testimonios sobre la Planta Nuclear*, April 1985, p.5.
[14] Ibid, p.6.
[15] Idem.
[16] A quasi-public corporation, the only producer and distributor of electrical energy in Puerto Rico.
[17] *El Imparcial*, 30 Aug 1972.
[18] Benjamín Ortiz, Industrial Mission letters.
[19] See Tomás Morales Cardona, 'Nuclear Energy Under Colonialism' in *Bulletin of the Puerto Rico Solidarity Committee* vol V, no.5, Nov/Dec, 1979, p.13.
[20] *Caribbean Business*, 5 Oct 1983, p.7.
[21] See Tomás Morales Cardona, *El origen de los problemas de la AFF a la luz del proceso decisional seguido con las centrales nucleares como alternativa energeítca para Puerto Rico* Jan 1979.
[22] See López, Meyn, op cit, p.140. The myopia reached such extremes that a superport was proposed to receive tankers bringing petroleum to Puerto Rico for refining. It was an absurd pretension to suppose that the island would become a supplier of raw materials for US industry.
[23] After the victory in Mayagüez, the communities had to confront a flood of proposals for energy plants promoted by a new weapon: a highly sophisticated publicity and public relations drive by the companies themselves. However, of the six projects, three have already been blocked successfully (Rincón, Yabucoa and Barceloneta) and the communities are continuing their struggle against the remaining three (Arecibo, Guyama, Guayanilla).

Green crime, green redemption: the environment and ecotourism in the Caribbean

Polly Pattullo

In the 1940s, the American war correspondent Martha Gellhorn, passed through the British Virgin Islands, and came across a cove which was 'a place where nothing had changed since time began, a half circle of white sand, flanked by huge squarish smooth rocks, the rocks overlapping to form cool caves and the water turquoise blue above the furrows of the sandy sea bed'.[1] Returning many years later, she found her cove 'full of sun-tanned bodies and ringed by boats, from swan yachts to rubber Zodiacs, and there were bottles and plastic debris on the sea-bed and picnic litter on the sand, for the rich are as disgusting as the poor in their carelessness of the natural world'.[2] The magic had become tainted.

Yet it is that Caribbean canvas, brushed blue for the sea and sky, green for vegetation and yellow for sand, and so conveniently splashed with hummingbirds and hibiscus, coconut palms and sunsets, that tourists have come to expect. Those well-edited images of the Caribbean environment are what tourists want; they go to the Caribbean for its climate, sea and beaches, not for its mountains and rivers, its cities or ruined battlements. Such demand has put its coastlines under enormous pressure.

The sort of tourism that now dominates the Caribbean, as Martha Gellhorn noted, has redefined its physical landscape. It has brought about the region's second invasion of land-snatchers; first it was the planters who changed the natural environment when they cleared the land for sugar-cane (islands now almost treeless, like Barbados and Antigua, were once shaggily forested). This time it has been the coastline which has been cleared. And it is the tourists, greedy for those rum-soaked images of the Caribbean, who are harvesting off the natural resources of land and water.

The coastal clearances have usually been along white sand beaches, on ancient and ground-down coral, predominantly on the west and sheltered littoral away from the rougher Atlantic shores. Large concrete hotels have been built close to the high-water mark, groynes and piers erected, marinas for yachts and deep-water harbours for cruise ships constructed. The great wetlands of the Caribbean have been grubbed out by developers eyeing their proximity to some of the region's best beaches. In Jamaica, Montego Bay's international airport was built on a wetland, while at nearby Ocho Rios, forty acres of swamp were turned into a resort with 4,000 beds and a cruise ship pier.[3] In a generation, the land and seascape have been transformed: the bays where once local fishermen pulled in their seine nets, where villagers went for a sea-bathe or where colonies of birds nested in mangrove stands now provide for the very different needs of tourists. The impact has been dramatic.

These transformations have been superimposed on a fragile environment particularly vulnerable to change. As Jean Holder warned in 1988: 'Our tourism product is our environment. We therefore destroy our environment at our economic peril.'[4] Yet, according to the environmentalists, serious damage has been done and continues to be done. According to Calvin Howell, Director of the Caribbean Conservation Association (CCA), 'We are a very fragile area and the environment is tourism's resource. However, there are countless examples throughout the region to suggest that there is a tendency to overlook the well-being of the environment in order to maximise the tourist dollar.' Significantly, he adds, 'it is hard to find examples of good practice in the region.'

Paradise lost

The catalogue of environmental destruction directly attributed to the growth of the tourist industry is long. It includes the erosion of beaches, the breakdown of coral reefs, marine and coastal pollution from watersports, the dumping of waste and the non-treatment of sewage, sand mining, the destruction of wetlands and salt ponds. In many cases, the impact is inter-related, locked into a chain of tourist development where short-term gain takes precedence over long-term protection. For example, a hotel cuts down coastal trees to improve the view from its bedrooms; this accelerates coastal erosion and sand loss; then, when a hotel builds a concrete jetty for its dive boat it loses even more sand because sand from the newly shaped beach is washed on to the coral reef. The result is two-fold: the sandy beach has become smaller and the marine environment has been spoiled. What the tourist came to enjoy no longer exists in its pristine condition.

For the Caribbean's smaller islands and communities, the greater the numbers of tourists the greater the pressure on the physical environment and the greater the demands on limited resources. Yet the major thrust is always to increase visitor arrivals.

At some point what is known as the 'carrying capacity' threshold of a tourist spot is reached. This is the point, according to the World Tourism Organization, 'when negative factors start to operate.' It is a vague enough definition, but it is at that point that the tourists vote with their feet and go elsewhere.

'However tolerant local inhabitants will be, it is clear that the tourist who has a choice, will not put up with litter, beach erosion, water pollution, dead coral reefs and other fall out from environmental neglect', Jean Holder told a London audience in 1993 in a paper on the compatibility of conservation and economic growth. He talked of the last phase of the model of tourism self-destruct: 'As the place sinks under the weight of social friction and solid waste, all tourists exit, leaving behind derelict tourism facilities, littered beaches and countryside, and a resident population that cannot return to its old way of life.'

Barbados, and in particular its south coast, has at times seemed to be closest to that mark (St Maarten looks in parts as if it should be). Even without tourists, Barbados is a densely populated island. In 1992 its population was just over a quarter of a million, the majority of whom live along the narrow coastal strips of the south and west coasts (roughly one-sixth of the island). These two coastlines are also the heartland of the tourist industry, adding some 400,000 stayover visitors and the same number of cruise passengers per year to the resident population.

The result has been overwhelming pressure on the tourist zones, especially along the south coast from Oistins in the south-east to Bridgetown. And despite a new sewage plant for the south and attempts to stem coastal erosion, tourists have moved away. Tourist arrivals started to show a decline first in the early 1980s and again from 1989 onwards. Yet the acting Minister of Tourism insisted in 1994 that 'there has been a negligible impact of environmental degradation. We have been committed from day one to quality tourism.'

It is not, however, just the cheaper end of the tourist market which poses a threat to the environment. In an attempt to attract the high-spending, up-market tourist, Barbados fell in love with real estate and golf, a game which not only claims agricultural land but uses 600,000 gallons of water per course per day. The Royal Westmoreland Golf and Country Club, a US$400 million residential and tourist resort, is the island's largest private investment. A 27-hole course was due for completion by 1996 despite the further strains on the environment.

While Barbados' 'mature' tourist industry has been forced to examine the reasons behind its environmental decline and address them with a still-on-paper regeneration project, the region as a whole pays lip-service to the environment, allowing its degradation to continue largely unchecked and unmanaged just so long as tourist arrival figures look good. Yet as the West Indian Commission's *Time for Action* put it: 'We cannot assume automatic victory in our battle against environmental degradation of our tourist destinations in the region.'[5]

Planners and politicians

In many instances, the institutions and mechanisms that are required to best prevent the region from destroying what it needs most are absent. Added to this are a lack of trained people, scientific technology, an educated public and, crucially, adequate financing.

Many of the reasons for the degradation of the physical environment can be traced to institutional weakness in the public sector, a problem that continues to plague the Caribbean's decision-making processes, and a lack of political will. As Calvin Howell of the CCA explains: 'The problem has been exacerbated by an attitude that approves of short-term gain rather than long-term sustainable development.' Or, as Klaus de Albuquerque, an American academic who has written extensively about the Caribbean, concluded (with particular reference to Antigua): 'The hard reality is that the majority of Caribbean governments are the worst regional environmental offenders, and even in the most liberal of democracies, the kinds of participatory planning processes necessary for sustainable utilization of resources, are often absent.'[6] A summary of a marine parks management training workshop in 1992 concluded that 'most of the Caribbean's marine protected areas are not being adequately managed'. It also emphasized that management is stymied by a lack of financial, human and infrastructural resources. 'Appropriate institutional and legislative structures, and more vigorous public support are also needed.'[7]

In Antigua, where some of the worst practices obtain, few management structures exist and the cavalier, if not corrupt, behaviour of politicians has meant that environmental protection legislation has been rendered almost meaningless. One development project at Coconut Hall was only thwarted by a handful of local people concerned about another attack on the island's coastline.

Environmental damage is often the result of the public sector's inability to impose careful control over tourism developments. But what has also been working against the 'greening' of the Caribbean is public perception. The region's elite has thrown out the old (associated with backwardness) for modernity and has often been responsible for destroying its own environment, egged on by external financial interests. At the same time, the poor have also damaged the environment in their struggle to survive: they have collected coral to sell, littered the beaches with their own waste and thrown their own garbage into the sea.

With tourism, land has become valued for its economic potential, and landscapes have been changed out of recognition by real estate and tourist facilities. Although land has a specific cultural, economic and spiritual role in Caribbean societies, traditionally it has not been 'appreciated' in a European sense. This view is, of course, changing as societies have become 'modern' and 'tourism awareness' has introduced the idea that natural beauty is to be admired for itself. Thus a perception develops that 'tourists will like that view.' Separately, the Rastafarians have also taken a lead in environmental awareness with their emphasis on the value of what is natural or 'irie', their knowledge of and care for the land, and their rejection of materialism. 'Better a piece of land than a big Cadillac,' according to a St Lucian Rasta elder.[8]

If the Rasta's words are to be heeded, one function of sound environmental management is the damage-limitation exercise to stop the decline of coasts and reefs, wetlands and rainforest. Another would be to put that understanding to good effect and to use tourism to protect the environment rather than to destroy it and so contribute to sustainable development.

Ecotourism

Governments and tourist establishments, locked into their old-fashioned sand-and-sea agenda, have been slow to tune in to this new thinking. However, by the end of the 1980s, interest was beginning to grow in what had become known as ecotourism (otherwise described as responsible, alternative, caring or green tourism).

An early definition of ecotourism was of 'travelling to relatively undisturbed or uncontaminated natural areas with the specific objective of studying, admiring and enjoying the scenery with its wild plants and animals, as well as any existing cultural manifestations (both past and present) found in these areas...'[9] This demand-led definition, however, omitted anything about the needs of developing countries. This has led to further definitions. This one is from the Ecotourism Society: 'Conserving natural environments and the well-being of local people through responsible travel,' a definition later amended to include the protection of historical and archaeological resources.

Ecotourism has usually come to mean small-scale, up-market tourism where visitors respect and express interest in local natural history and culture. Compared with mass tourism, ecotourism also supports a larger degree of local involvement, better linkages, a reduction in leakages and increased financial returns leading to sustainable development. Jean Holder of the CTO has also described ecotourism as a tool for learning. Addressing the first Caribbean Ecotourism Conference in Belize in 1991, Holder said that ecotourism could present the last chance for the region:

> to find the formula which does not at one and the same time entice the visitor, while alienating the local residents. All of our tourists may never be ecotourists. But even those, who come primarily to laze on our beaches, can by the provision of creative programmes, be interested to participate in an activity which teaches them a great deal more about us and our country and, ultimately, about themselves.[10]

Alarm at the damage tourism has already inflicted on the environment may have been one reason why the Caribbean began to talk ecotourism. Another was the trend in North America and Europe for 'green' holidays, away from sun-bathing and duty-free shopping. Figures showed that by the beginning of the 1990s nature-related travel was the fastest-growing sector in international tourism. According to the World Wildlife Fund, about fifteen per cent of the world's 450 million travellers in 1991 were taking hiking shoes and rucksacks on their holidays rather than sarongs and swimsuits.

Much lip-service is now paid to ecotourism in the Caribbean. It is seen as 'a good thing' by some (or dismissed as a fad or a meaningless buzzword by others). Even Barbados, a long-time exponent of traditional beach tourism and with few 'undeveloped' wild places left, except its splendid east coast, has expressed an interest in ecotourism. The acting Minister of Tourism was in 1994 all for ecotourism: 'We have trails being developed; we can create things to make the hikes interesting and people don't have to go for miles to get there.' The Minister's assumption appeared to embrace the notion that any activity involving a tree rather than a beach was a credible stab at 'ecotourism'.

Many Caribbean countries now promote ecotourism to describe any part of their 'tourist product' that focuses on natural attractions. Responding to demand, they have introduced ecotourism as an extra dimension to standard beachside holidays. Grenada, for instance, opened its 450-acre Levera-Bathway National Park in 1994. The then Tourism Minister, Tillman Thomas, said that the park would protect the endangered turtles which nest in the area and simultaneously aid Grenada's economic base. 'Because of the resources here, we have a great potential for developing ecotourism, a product that is in harmony with the environment'.[11]

On a much larger scale, Jamaica's Blue Mountain/John Crow Mountain National Park, the country's first national park, was opened in 1989 to help stop deforestation, which had caused serious soil erosion. The 200,000-acre park has developed a range of recreational and educational activities for tourists and Jamaicans, which in turn will provide employment for local people. The concept behind the park is for Jamaicans to adopt sustainable land-use policies while preserving the area from detrimental development. In 1992, the Blue Mountain/John Crow National Park and Montego Bay Marine Park received a US$100,000 'debt for nature' grant from the Puerto Rican Conservation Trust.

At Jack's Hill, a community on the edge of the national park, a special conservation area is being managed by local people to promote ecotourism and sustainable agriculture in the Blue Mountains. Tourists who stay at Maya Lodge, the headquarters of the Jamaican Alternative Tourism, Camping and Hiking Association, use the Blue Mountains for various sporting activities and also spend time with local farmers and residents. Maya Lodge is also a model demonstration site for a community reforestation and environmental education programme and provides support, research and training for other ecotourism operations.

While the Blue Mountain National Park integrates local and tourist needs, the creation of the Virgin Islands National Park on St John's, established under the control of the US Federal Government in 1956, illustrates how conservation can create conflict between local needs and conservationists schooled in North American perceptions about natural wildernesses. Firstly, the landscape of the park, which reverted to 'nature', alienated the local population, who had been used to cultivating the land; secondly, economic benefit to St Johnians was, according to one study, limited. 'The park service has not sought to stimulate local business outside the park, but rather seems to have circumvented it whenever possible,' reported an anthropologist in 1980.[12] Another problem is that ecotourism on St John's is led by American-style and expensive innovations. At Maho Bay Camp, new villas, made from recycled materials, each have a personal computer, tracking and controlling energy and water consumption.

Unlike the Virgin Islands National Park, most national parks in the Caribbean are often dependent on voluntary contributions or funding by outside agencies, such as the World Wildlife Fund. Both the El Yunque Tropical Rain Forest in Puerto Rico, a 28,000-acre bird sanctuary with a rich and rare variety of trees, and the Asa Wright Nature Reserve in Trinidad, which attracts naturalists and ornithologists, depend in part on visitors' receipts. Again, many of the most important protected sites in Belize, such as the Half Moon Caye, the Cockscomb Basin Wildlife Sanctuary and the Community Baboon Sanctuary, are not managed by the Belize government but depend on the voluntary Belize Audubon Society for funding and management. The costs of running sanctuaries and maintaining national parks are high, but introducing entrance fees and merchandising sales create further management problems.

A distinction, however, must be made between those countries such as Grenada and Jamaica which tie tourism in with sustainable environmental projects while at the same time pursuing traditional tourism, and those which claim that ecotourism defines the shape and strategy of their tourist industry. Belize and Dominica fall into the latter category.

The attractions of ecotourism as a tourist model for both Belize and Dominica was that both countries have stunning landscapes, flora and fauna and marine environments. Both had avoided mass-tourism development by default because of poor communications, both internal and with the outside world, poor infrastructure, and, in the case of Dominica, because of a lack of white sand beaches. However, both these predominantly agricultural countries have turned to ecotourism for the classic reasons: to diversify the economy (in Dominica's case to offer an alternative to bananas), to generate foreign exchange and to provide jobs.

Ecotourism has become a central platform of tourist development in Belize, Dominica and, most recently, Guyana, while being incorporated as an alternative 'niche' within the more mass markets of other destinations. It has provided a sort of lifeline, but, it is not without its difficulties. While ecotourism could conserve the environment, it also makes demands on it, and while it can offer sustainable development and integrate local people, it can sometimes also alienate them almost as easily as mass tourism.

As Erlet Cater, a British geographer and commentator on Third World ecotourism has written: 'There is a real danger that ecotourism may merely replicate the economic, social and physical problems already associated with conventional tourism. The only difference... is that often previously undeveloped areas, with delicately balanced physical and cultural environments, are being brought into the locus of international tourism.'[13] At that point, ecotourism in the Caribbean, or indeed anywhere else, no longer has any specific meaning.

Yet the environment of the Caribbean needs to remain the centrepiece of the region's enchantment. Firstly, it must do so to provide a sustainable future for its people and secondly to fulfill the fantasies of those millions of visitors who, as they leave the plane, take their first sniff of that still sweet Caribbean air.

* Extracts from Polly Pattullo, *Last Resorts: the Cost of Tourism in the Caribbean*, Latin America Bureau, London, 1996.

[1] Martha Gellhorn, *Travels With Myself and Another*, London, 1983, p.70.
[2] *Ibid.*, p.107.
[3] Peter Bacon, 'Use of Wetlands for Tourism in the Insular Caribbean', *Annals of Tourism Research*, vol 14, 1987, pp.104-117.
[4] Jean Holder, 'Tourism and Environmental Planning: An Irrevocable Commitment', *Caribinia*, Caribbean Conservation Association, *1988.*
[5] West Indian Commission, *Time for Action: Overview of the Report of the West Indian Commission*, Barbados, 1992, p. 107.
[6] Klaus de Albuquerque, 'Conflicting Claims on Antigua Coastal Resources: the case of the McKinnons and Jolly Hill Salt Ponds' in Norman Girvan and David Simmons (eds), *Caribbean Ecology and Economics,* Barbados, 1991.
[7] *Caribbean Conservation News*, Barbados, vol 6, no 1, March 1993.
[8] Yves Renard, 'Perceptions of the Environment', Caribbean Conservation Association, Barbados, 1992, p.53.
[9] Vera Ann Brereton, *Eco-Tourism in the Caribbean*, Caribbean Hotel Association Handbook, 1993.
[10] Jean Holder, paper delivered at first Caribbean ecotourism conference, Belize, 1991.
[11] Caribbean News Agency (CANA), 16 May 1994.
[12] Karen Fog Olwig, 'National Parks, Tourism and Local Development: A West Indian Case.' *Human Organisation*, vol 39, no 1, 1980.
[13] Erlet Cater, *op. cit.*, p.19.

Where will all the garbage go? Tourism, politics, and the environment in Barbados

Hilary McD. Beckles

The leading national newspaper voted it the 'Controversy of the Year'. What a stink, Barbadians thought, as they settled down with thousands of tourists to the 1995 Christmas celebrations. Since 3 February, controversy had erupted in towns, villages and Parliament when the government announced its intention to close a badly managed waste disposal landfill located at Mangrove in the centre of the island and establish a new facility in Greenland, a village embraced by the beautiful Scotland District near the Atlantic coast.

For twenty years the Scotland District, on account of its natural vegetation and spectacular scenery, has been designated a national park. Greenland is located at the centre of the park, and residents there, mostly peasants and estate labourers, have become accustomed to car loads of tourists, scout troops, 'green' hikers, birdwatchers and lovers trekking through their backyards. Villagers were shocked when their Prime Minister revealed the news that within six months garbage trucks would be added to this traffic, as the national waste disposal landfill was to take up residence in their green lands.

In a subsequent statement delivered by the Minister for Health and Environment, Liz Thompson, citizens heard that the 'decision was reached after long and careful thought and after examination of all the possible sites for solid waste facilities'. The Minister assured everyone that in reaching the decision; 'the Government has sought and relied on the advice and guidance of both international and local engineers and local experts.' The landfill site at Mangrove, the Minister said, had 'become a national problem which is costing the Government $40 000 a day to correct'. Some aspects of the problem, however, could not be measured in financial terms.

In recent years the mismanagement of the Mangrove landfill had become a major scandal. Neighbouring communities had paid an enormous price through their exposure to polluted air, the foul smell of which caused motorists to seek alternative routes. The residents of nearby Arch Hall Village, downwind of the site, were not so fortunate. Children broke out in rashes, and adults complained of eye and throat irritations. As the pile of garbage mounted, high winds took debris and odour to the air and deposited them within households. Residents named the site Mount Stinkeroo, and demanded its closure. On a weekly basis, dozens of cruiseships not only deposited tourists and precious foreign currencies on the island but also off-loaded tons of garbage that ended up in Mangrove. Environmentalist Dr. Mark Griffith noted that the site was absorbing thirty per cent more than it was designed to take. 'Tourism is our business, let's all play our part', went the slogan on the government-sponsored 'infomercial' broadcast daily on TV. Meanwhile the ugly side of tourism continued to be dumped at Mangrove.

During the 1993-94 election campaign, David Simmons, a parliamentarian for the area and senior opposition spokesman, vowed to his constituents and promised the country that if his party were returned to government, Mangrove, alias Mount Stinkeroo, would be closed and residents whose health had been adversely affected would be compensated by the government. He accused the government of criminal negligence and indicated his desire to take it to court to seek damages for his constituents. The government, he said, had also failed the country in not formulating and implementing a rational garbage disposal policy while spending millions of dollars on the promotion of heritage and environmental tourism. The double edged sword was clearly exposed.

A site is chosen

When the opposition Barbados Labour Party gained the majority of seats and formed the new government, David Simmons was sworn in as Attorney General. He immediately committed his party to the closure of the landfill at Mangrove, vowing to find an alternative site and compensate affected residents. But these decisions merely highlighted the lack of policy on environmental protection in a country dependent on tourism. Taking the waste elsewhere was not a solution. Stories of hotels dumping liquid waste in the sea, for example, remain as disturbing as the development at Mangrove. Pronouncements by the government promoting eco-friendly tourism products appear contradicted by infrastructural underdevelopment and its own waste disposal practices.

When it was announced that the Scotland District national park was to provide the alternative landfill site, Minister Thompson asserted that given the size of Barbados, 'it makes sense to choose a site which offers the possibility of a long-term operation'. The national park takes up fifteen per cent of the island's total land mass and Greenland comprises 74 acres of land. The following features of Greenland were deemed appropriate for a landfill:

1) It is composed of clay soil and clay will reduce the cost of constructing and operating the landfill.

2) Greenland is a valley, in fact it is a sand/clay quarry site, reducing the possibility of odour dispersion.

3) The site is vast; it is off the beaten track; it is screened by trees and is not visible from any of the nearby tourist attractions.

4) It is downwind of residential development.

Government invited residents and 'other interested persons' to discuss the decision and to hear 'how Greenland will be managed so as to prevent environmental problems'.

The public fights back

Within days of his announcement public debate swept the island like a hurricane; political clashes, threats of physical abuse and community outrage dominated news, radio talk shows, current affairs programmes, editorials, opinion columns, and letter pages in the press. Greenland residents, and their urban 'green' supporters, took to the streets in protest, and threatened to 'sleep in the roads' before garbage trucks were allowed in their community. David Seale, a well-known businessman, amazed by the magnitude of public protest against the government he supports, stated sarcastically: 'outside of political election campaigns I have never seen so many arguments mustered, or invented, for a single cause. Never before did I know Barbados boasted so many environmental experts or 'green-people.' It was certainly the first occasion that an environmental issue had assumed centre stage in national politics.

The 'green coalition' that came out in support of Greenland residents included persons from the small but active Barbados Conservation Society, the well-established Barbados National Trust, the Barbados Agricultural Society and students and individuals involved in a range of direct tourism services. By far the largest single group, however, comprised individuals who took a 'common sense' objection to the introduction of pollutants into the national park that functioned for them and for tourists as a place to get away from the suburban congestion of the heavily overpopulated island. With 257,000 inhabitants and an equal number of tourists competing simultaneously for the use of an island 166 sq. miles in size, a premium would be inevitably be placed by all on the serene atmosphere of the Scotland District.

Residents vowed to march on Parliament, and to pay a visit to the Governor General in order to register their objection. A Greenland Protection Group (GPG) was established under the leadership of Richard Goddard, a well-known white farmer with a business interest in the area, and cousin to the Minister of International Transport and Business. At a public meeting held under a starry sky, Goddard warned the government: 'We want to give notice now that although we may be small in number, the pen is mightier than the sword and if Government announces that the landfill will come to Greenland, then the pen will start to work.' He argued that the landfill was incompatible with both a national park and government policy to promote private sector eco-friendly tourism.

For decades, residents in the parish of St Andrew, where Greenland is located, have been complaining of infrastructural and economic neglect by successive governments. Conscious of this history of marginalization, they had good cause to consider themselves 'dumped' on by an urban-based political process that sees little political mileage in environmental conservation and ecological issues. They were quick to react, and within a short time had succeeded in mobilizing considerable public support and intellectual expertise in their cause. Speaking on their behalf, a foremost expert on Barbados' geology, Dr Robert Speed, of Northwestern University, denounced the government's decision as 'high risk' on account of sub-soil instability in the area. Christine Keller, an international project analyst, stated that a detailed costing would indicate the comparative disadvantage of Greenland as a site. Parliamentary representative for the St Andrew parish, George Payne, supported constituents against Cabinet colleagues, and stated publicly that 'it is environmental madness to place the dump in the middle of a national park'.

Under pressure from a well-organized and very effective community action programme, Cabinet agreed to defer the start of construction at Greenland until an Environmental Impact Assessment (EIA) and Feasibility Studies were carried out. According to a memorandum circulated to senior staff in the Ministry of Health, these studies were to be completed by July 1995. Opposition to the landfill mounted. A consultant report conducted by Stanley Associates Engineering Ltd., R.W. Beck & Associates and Consulting Engineering Partnership Ltd., indicated that contrary to government claims, Greenland was the least suited of all four landfill sites being considered. Even Chief Town Planner, Lionel Nurse, disagreed with the choice of Greenland (for many years, Nurse served as a board member of the Scotland District Authority). The minister, however, was adamant, and declared that a defeat on this issue could very well bring down the government.

The GPG, a collective of residents, technical experts, and environmental activists, seemed as determined. Goddard, its leader, declared his intention to petition the Inter-American Development Bank (IADB), an expected funding source for the projects, in order to demonstrate the folly of government's position. Goddard and Minister Thompson met face to face on a televised public affairs programme, and in addition to verbally denigrating each other, outraged viewers with the intensity of their personal animosity. Goddard proclaimed that the minister was 'hell bent' on placing Barbados in the Guinness Book of Records for being the first country to place a national dump in a national park. Minister Thompson, seeking to discredit Goddard, described him as a disrespecting 'Causcasian' male with a problem regarding black people in government. Despite the fact that Minister of Labour, George Payne, had aggressively attacked his cabinet colleagues, and Minister Thompson had received public condemnation for an alleged 'racist' attack on Goddard, Prime Minister Arthur informed the media that the Cabinet jobs of both ministers 'were safe'.

Government counter-attack

The publication of the draft report of the Environmental Impact Assessment in August 1995 opened new controversies, and intensified public debates in the media and the Greenland community. The government claimed that the report was not a legal requirement but one stipulated by potential funders, such as the IADB and the Caribbean Development Bank. Moreover, the EIA was not intended to offer a position on the Greenland site but to provide information on the advantages and disadvantages of the site in order to guide policymakers.

The findings of the EIA report were considered ambiguous, providing important evidence for both sides of the debate. It did indicate, however, that Greenland would cost US$10 million more to run than Mangrove and that Mangrove seemed more 'environmentally and socially acceptable'. At a public meeting held in September, 125 people heard that their presence was designed as a formality to satisfy the basic requiremnts of the IADB that there be 'public consultation'. Only a single copy of the 1,000-page report was made available to the entire population of 257,000 who were invited to read, digest, and comment upon its contents within five working days.

Environmentalists and Greenland residents continued their protest, while Professor Speed predicted dire consequences if the dump were to be cited at Greenland. Respected soil consevationalist, Edward Cumberbatch, pointed out the inconsistencies in the sub-strata, and the instability of the Greenland area due to its drift characteristics, located as it is in an area with annual rainfall in excess of 60 inches. By the end of September, most of the reasons proffered by government for siting a dump at Greenland were effectively contradicted by real experts. But despite the perceived water and geological problems, local expert advice, and the very heavy ancillary cost associated with the Greenland project, it appeared that the government's mind was set.

When the heavy rains of August produced no signs of flooding at the site, the government was even more resolved. 'We have no problems here,' declared Dr Hugh Sealy, Deputy Project Manager of the Sewerage and Solid Waste Project Unit. Greenland, he said, had held firm while the Mangrove Landfill was heavily flooded. Data compiled by the Soil Conservation Unit showed that rainfall at both sites was comparable; 181.5 mm (7.14 inches) at Greenland, and 240mm (9.45 inches) at Mangrove. In mid-September the press reported that the IADB had agreed to fund the controversial landfill which was estimated to be operational for about twenty years. The press also reported that the IADB had given its consultants (Stanley Associates) the green light to start designing the facility at Greenland, which is expected to include an area for disposal of hazardous waste. These reports prompted environmentalist, Hermon Lowe, to inform the press: 'I am against it; it is right smack in the centre of the national park. Why go defacing your front yard when it's becoming the breadbasket for tourism?'

Ally Terry, a journalist with the Nation Newspaper, who had been covering the story with great diligence, reported:

> Not content with luring visitors to its shores with luxury all-inclusives, golf courses, and maybe even casinos, Barbados has also been actively marketing itself as an ecotourism destination. But being the overcrowded, pancake-shaped dot in the Caribbean that it is, can Barbados realistically compete with say, Dominica, with its mountains and Guyana, with its river? Barbados has the rugged East Coast, Harrison's Cave, Turners Hall Woods, and Greenland; it is crucial that the few natural, unspoilt acres that Barbados still boasts, should remain...

Following these developments, the GPG organized a series of public meetings and petitioned the IADB. Opposition political parties took to the village in a full show of solidarity with residents. Under siege from gathering forces, the government announced a 'seven-month stay of execution'.

Victory for the protesters?

By the end of 1995, the issue was still unresolved. 'If you look between the lines,' says naturalist, Dr Hudson, 'there seems to be a diplomatic thumbs down for Greenland and thumbs up for the original proposal of just making a good job of Mangrove.' A new commencement date for the Greenland site has been set for 1996, but no specifics have been advanced, while a vote by the IADB's Board of Directors for a loan for the project has been rescheduled to sometime in the second quarter of 1996, on the grounds that the project is controversial and that there needed to be a 'meeting of minds' on the issues.

The thinking in the Greenland community is that it had won round one of the contest with the government. One resident is reported by the press as saying that any time the minister tries to turn their community into a dump 'she goin' get another fight'. The minister backed down, they believe, because 'she never thought she would get such a fight, because poeple always think that we in St Andrew are dumb, but we ain't'.

The crisis is indicative of a deep-rooted indifference to the environment, an attitude that has characterized the history of political and economic exploitation of the West Indies. The culture of plantation management, which continues to prevent the majority from owning land in the countryside, has alienated people from environmental issues. The state and private sector organizations are currently struggling to inform their judgements with an eco-sensibility, but at best their actions are of a minimalist nature and designed to pacify popular opinion rather than to promote environmental awareness. Neither rapid suburban expansion in recent decades nor the shift from an agricultural to a tourism services economy in the 1960s and 1970s has been accompanied by any meaningful environmental impact assessments. Damage control continues to dominate public sector responses to these developments, and the Greenland crisis is just the latest manifestation of official insensitivity to environmental concerns. The government is now sufficiently sensitized to reflect critically upon this tradition of ecological neglect, and to take on board the sentiments expressed by protesters. At last there is an opportunity to look at the future of the island through 'green' lenses, and to develop a policy trajectory that is based upon a less destructive relationship between people and land.

5
URBAN ECOWARRIORS: ENVIRONMENTAL CONFLICTS AND INITIATIVES IN LATIN AMERICAN CITIES

Enlightened cities: the urban environment in Latin America

Julio D. Dávila

The world is rapidly becoming urban, and Latin America is a vigorous contributor to this. Between 1960 and 1995, the number of city-dwellers in the region rose by over 200 million, thus adding one in every seven new urbanites to the world's population. Today, three out of four Latin Americans live in cities or places that may be classified as urban – much the same as in Europe. By the end of the 1990s, Latin America will boast four of the world's twenty 'mega-cities' – each with a population of over ten million inhabitants. At the same time as they have opted for urban life and urban jobs, the inhabitants of Latin America and the Caribbean have also become wealthier (on average), albeit at a slower pace than their counterparts in the more successful economies of southeast Asia. But increased living standards have not removed serious income disparities which remain one of the greatest threats to economic prosperity and environmental sustainability in the region.

For a tourist returning to one of the region's capital cities after an absence of thirty years, the most visible signs of change will be in the astonishing number of cars, taxis and buses that clog the busier streets and choke the population in clouds of smoke. For the visitor that ventures away from the historic areas and the lavish, American-style shopping malls into the poorest *barrios* or *favelas*, even more striking might be the lack of adequate sewage or the erratic supplies of drinking water, the unpaved streets and even the tensions often leading to violence, the results of marked social contrasts. But were they to sit down for a beer or a *mate* tea with members of these communities, they would be very likely to discover an exciting wealth of shared experiences, of joint efforts – often with the guidance of young professionals and the support of the mayor – to protect their neighbourhood against regular flooding from a nearby river, plant trees, recycle their garbage or even provide jobs for single mothers. This would give them a more accurate vision of today's urban environment in Latin America.

Latin America: an urbanising region

In 1940 the majority of the population of Latin America and the Caribbean lived in the countryside (with a few exceptions such as Argentina and Chile). Fifty years later the situation had completely reversed. Initially, this was largely due to the combined decisions of millions of individuals to move to the city in search of jobs, better access to health care and education for their children than their village or town could ever provide. In some cases they were spurred by the despair of living off a land that did not belong to them. For others, their jobs had become redundant by the introduction of agricultural machinery which could do their work faster, more cheaply. Migrants were among the youngest, best educated in their communities. Some went to their region's capital, others headed straight to the national capital, which thus benefited from welcoming the best minds and most able hands.

For decades, the vast majority of migrants were proved right. Rising incomes, along with improved and expanded health care and education, increased living standards, falling mortality rates and a growing life expectancy, were the trademark of economic growth throughout Latin America up to 1980. In an economic environment which was rapidly changing, where traditional or extensive agricultural farming methods were no longer able to keep up with a much increased demand from both the national and the international markets, mechanised agriculture, manufacturing and city-based services appeared as the more promising areas for expansion.

Despite the alarm of ruling elites who (like their European counterparts centuries earlier) saw the potential threat of large numbers of concentrated people, it all made economic and environmental sense: urbanization is closely linked to economic growth, as cities are the privileged location for more highly productive activities and are therefore able to sustain much larger numbers of people with higher incomes in much smaller areas. Such is the case with São Paulo which generated 48 per cent of Brazil's industrial output in 1970, while having less than nine per cent of the country's population; or Santo Domingo, with one fifth of the Dominican Republic's population and where, in 1981, 70 per cent of the country's banking transactions took place; and also of Guayaquil which produces a third of Ecuador's yearly output and yet has a mere 15 per cent of its population.[1] An equally important advantage of concentrations of population is the fact that it makes the provision of services and infrastructure comparatively less costly in per capita terms, thus reaching greater numbers.

By 1960 there were already more people living in urban than in rural areas. By 1995, the region had changed beyond recognition: not only did the urban population exceed the rural population by a factor of three, but there were fewer people living in the countryside than in 1960. Urban growth was particularly impressive in the largest cities which were rapidly engulfing nearby villages and towns to form metropolitan areas. Between 1960 and 1970 the population of Mexico City nearly doubled, as over 1,000 new people appeared daily in its streets.[2] Many smaller cities and towns also grew very fast, as mining and frontier areas were opened to commercial exploitation and areas of traditional exports boomed. Valledupar, in the heart of a cotton and cattle ranching region in northern Colombia, expanded at a hair-raising 12.9 per cent annually between 1951 and 1964.

In the 1960s and 1970s, national and city governments could barely keep up with the demands for services and infrastructure that these inconceivable volumes of people and new activity required. It was an expensive affair: it cost Medellín some US$1,200 million to build an elevated rail transport system, an average of US$580 for each of the city's inhabitants, or nearly half their annual per capita income; every kilometre of Caracas' metro cost Venezuelan taxpayers US$150 million.[3] While most of the new national wealth was now being created in cities, city governments could rarely muster enough financial or human resources to supply basic services like water and sanitation to ensure adequate coverage to firms and residents alike in a land area which expanded sometimes faster than the population.[4] The poor suffered most in the process. Under military regimes or unrepresentative governments, voices of protest from informal settlements were often repressed or dismissed by ruling elites, oblivious of the discomfort of packed commuter buses and suburban trains.

Few countries escaped the brutal double blow of a mounting burden of national debt and the world recession of the early 1980s. The fast rise in interest rates and the parallel drop in the price of exports of raw materials combined to stifle the growth of national economies. The rapid growth in *per capita* incomes of 2.5 per cent per year in the 1960s and 3.1 per cent in the 1970s now gave way to falling incomes at an average of 0.5 per cent in the 1980s.[5] By 1990, the mean income of a Peruvian and a Nicaraguan was a third lower than in 1980, for example.[6]

Manufacturing industries and services which had flourished under the shelter of decades of protectionist policies and low interest rates suddenly collapsed. Cities and their inhabitants suffered badly. Between 1973 and 1984 nearly 68,000 manufacturing jobs were lost in the metropolitan region of Buenos Aires alone.[7] Manufacturing firms also left São Paulo and Mexico City, as rising costs of land and labour on top of more stringent pollution controls made smaller cities within commuting distance more attractive to investors. As people used their ingenuity to survive, 'informal' (ie. statistically unrecorded) activities flourished. The cost of debts contracted in earlier decades, a reluctance to raise property taxes and the inability to charge taxes from unrecorded activities combined to throw many municipal governments into a serious crisis. In many cities, the results became visible in poorly maintained infrastructure and services, the incapacity to undertake new public investment and sometimes in large numbers of unpaid municipal employees.

Structural reforms introduced in the 1980s have prompted governments to reduce subsidies on foodstuffs, transport, health and other urban services. The resulting higher prices and lower earnings have hit poor and middle income households particularly hard,[8] contributing to a level of urban poverty not witnessed in the previous decades of rapid growth. The numbers of Latin Americans earning less than US$60 per month, and thus classified as living in poverty, rose to an unprecedented 200 million in 1990, nearly half the continent's total population. Considering that the rural population had contracted and the proportion of rural poor had remained stable, these figures reflect a steady rise in the numbers of urban poor. Poverty is now firmly on the urban agenda.

The challenges of a deteriorating urban environment

A deterioration in the urban environment is neither irreversible nor an inevitable consequence of urbanization and economic growth, as the experience of the richer nations has shown. Between 1970 and 1988, when the economy of OECD countries grew by around 80 per cent, urban air quality improved substantially: particulate emissions dropped by 60 per cent and sulphur dioxides by 35 per cent; lead emissions fell by 85 per cent in North America and by 50 per cent in most West European countries.[9]

By and large, city governments in Latin America have managed to cope surprisingly well with the challenge of fast growth, by providing many of the services and amenities that rapidly growing populations and businesses needed.[10] Nevertheless Latin American urbanites today face enormous environmental challenges, a consequence of decades of population growth, economic expansion and an eschewed income distribution. Since no two cities are the same, the environmental problems of any city are unique – and, in Latin America, they are rarely known, as environmental data are still relatively rare, particularly for medium and small-sized cities. However, one may identify a range of issues found throughout the region, in some cases more critical than others, regarding air and water quality, supply of basic services, land scarcity and use, and sources and control of pollution.

As city economies expanded in the 1950s and 1960s, so did the number, concentration and diversity of polluting activities, such as manufacturing industries, vehicles and households. Caught unawares by the swiftness of the process, governments were often more interested in supplying an increasingly vociferous urban population and their employers with basic services than in confronting pollution. Any trace of life in many rivers in or near cities has long disappeared, following the dumping of toxic chemicals, untreated waste and uncollected rubbish.

So filthy are their waterways that, in the same way that Londoners shunned the Thames in decades past, many cities have turned their backs on them, and in the process on the poorer and less vocal inhabitants of towns and cities downstream. The dozen or so tributaries of Buenos Aires' River Plate are so bad that some, like the Morón creek, with a BOD concentration of 700 ppm,[11] are best classified as open sewers.[12] The longer solutions are deferred, the more expensive it becomes to deal with problems; relatively modest measures to clean parts of the Bogotá river, including a waste treatment plant at the worst point of industrial pollution and a waste water collector running parallel to the river, are estimated to cost over US$1 billion.

Another latecomer to the political agenda is air quality. Plans to seriously tackle Mexico City's rapidly deteriorating urban air, for example, were only drawn up in the late 1980s – though systematic monitoring started in the mid-1960s. The city's 2.5 million vehicles are the main source of air pollutants such as sulphur dioxide (associated with lung diseases), lead emissions and carbon monoxide (both of which may affect the central nervous system), nitrogen dioxide and ozone (responsible for mucus irritation and respiratory illnesses). By contrast, vehicles are not the main source of suspended air particulates; here, erosion brought about by deforestation and industrial emissions take the lead.[13] Tolerable limits on air-borne emissions were exceeded in Mexico City for over 250 days every year in the late 1980s, and air quality in many other large cities like Rio de Janeiro, Buenos Aires, São Paulo, Santiago, Lima and Bogotá has also deteriorated rapidly.

Since then, a combination of stricter controls on sources of pollution and the gradual loss of manufacturing industry have led to an improvement in air quality in some of the largest urban agglomerations. Some relief may also be found in the fact that poor air quality is not such an acute problem in smaller cities, simply because they have smaller populations and usually lower ratios of vehicles to population. With a population of half a million people and 100,000 vehicles, El Alto, Bolivia's fastest growing city located near La Paz airport, is not reported to have a serious air quality problem.[14]

Under the pressures of rapid population growth, for some decades many city governments placed an expansion of the supply of basic services to households and producers at the centre of their action. And yet the results are mixed. Even today, while the wealthier residents in most cities will generally have relatively reliable access to potable water, sewerage and electricity, the homes of many among the poor will lack one or more of these most basic of services, let alone telephone lines or rubbish collection. In 1990, of the poorest ten per cent of Lima's population only six in ten had access to water or sewerage in their home, while two in ten had to buy it (sometimes paying an extortionate 17 times more per litre) from water vendors; by contrast, over nine out of ten of the richest ten per cent of the city's population had access to both services.[15] While other large cities may have better coverage of basic services (especially among the richer groups, where full servicing is not uncommon), these figures help pinpoint a very important set of environmental problems found throughout Latin America.

A lack of drinking water and sanitation is the cause of a range of health ailments suffered by vast numbers of urban dwellers. An estimated 100 million people in urban Latin America live in areas which are inadequately covered by one of these services or lack access to adequate health services.[16] This is often coupled with overcrowding and poor housing conditions, where disease vectors can spread more easily and children and adults alike can be injured from household accidents such as fires or burns from cooking oils. These conditions mean that lower income groups are more likely to be ill or injured and to lose more work days – and hence income – as a result. The human and economic consequences of a poor environment can be devastating. It is estimated that in 1990 each person in Latin America and the Caribbean lost an average of 85 days as a result of diseases or injuries acquired during that year alone, with infectious or parasitic diseases accounting for 21 days lost. By comparison, only 43 days were lost in the richer nations, most (33.6) from non-communicable diseases such as cancer and heart problems.[17]

Although serious service deficiencies still persist in some neighbourhoods of national capitals, these are now more deeply felt in the rapidly growing peripheral areas of large cities or in smaller, fast-expanding, commuter towns close to large metropolitan areas, mining towns or dormitory areas of industrial towns, as municipal governments there often lack the human, legal and financial resources, as well as the political clout, to react to a rapidly changing situation.

Along with water, the lack of suitable land for urban use perhaps poses the greatest challenge for urban planners and local authorities. After all, without land, cities cannot grow. But it is land that makes or breaks the fortunes of politicians and speculators and is therefore potentially the most explosive issue. The best located areas – in relation to jobs, facilities and, increasingly, environmental amenities – will always command premium prices. This is why the urban poor in the Third World are regularly confined to distant locations, or to areas subject to heavy pollution, or to flooding and landslides. These phenomena are often wrongly termed 'natural' disasters, when in reality their causes are man-made, arising from social inequalities and poor land management. In cities with steep hillsides such as Caracas, Rio de Janeiro and Medellín, frequent landslides kill hundreds every year and thousands are left homeless. And it is because of their poor location that the urban poor suffer more from industrial accidents (such as Mexico City's 1984 gas explosion which killed 500) and even earthquakes (such as San Salvador's in 1986 which left a death toll of 2,000).[18]

The growing demand for urban open space, including parks and recreation facilities, pale next to these more urgent problems. Parks figured high in city master plans in earlier decades; but these elegant plans, inspired by the alien ideals of post-war Europe and the US, were overtaken by the overwhelming reality of rapid growth and a lack of resources even before they were finished. And yet open space is crucial for the well-being of a population subjected to a heavy daily dose of air and noise pollution, traffic congestion, social and work-related stress, increased violence and, in the present climate of reforms and liberalization, economic uncertainty.

It is only recently that the supply of open space has been resurrected as a priority by local governments. The problem is best tackled at the beginning of a city's growth, before land becomes impossibly expensive. But few city governments have the vision and the funds to follow the example of Curitiba, Brazil's 'environmental capital', where in twenty years of enlightened management, average open space per person rose from 0.5 to 52 m^2. By contrast, between 1979 and 1988, São Paulo's authorities struggled to add an extra 24 million m^2 of public garden area to the metropolitan region, thus raising the average of open space per person to a mere 4.5 m^2.

Waste seems to be an environmental problem about which there is a large measure of agreement in Latin America. Like inadequate water supplies, poor solid waste management poses serious health problems for rich and poor alike. For decades, a central concern of governments and urban dwellers has been simply to take rubbish away from places where it poses a health hazard. But pressure is mounting for a more rigorous approach to waste, while there is growing public interest in the social and environmental aspects of waste disposal and recycling.

This is not surprising, for two main reasons. Firstly, not all the waste produced is collected; in the late 1980s, one third of Guatemala City's garbage and one-fifth of Bogotá's was discharged in clandestine dumps or left to rot, simply because the authorities could not cope with it. In Buenos Aires in 1989, 16 per cent of the poorest households had no collection service whatsoever so they had to resort to burning the rubbish locally (thus contributing to air pollution) or dump it illegally. Few cities could tell a different story. Secondly, finding an area to dispose of solid waste has become a major headache for urban managers: no official landfills have opened in Greater Buenos Aires for a decade because no local authority will allow the location of one within its jurisdiction; however, over a hundred illegal sites (along with informal and usually non-technical disposal systems) have opened to cater for growing volumes of rubbish.[19]

Although much of the rubbish produced by Latin American urban households is of a vegetable – and therefore perishable – nature, it is estimated that between 10 and 25 per cent of household rubbish could be recycled, compared to as much as 75 per cent in Europe (although at present recycling as practised in Holland and Germany is closer to 50 per cent). This is particularly true of cardboard, paper, scrap iron, and glass which have been collected and sold by recyclers for decades in most Latin American cities (see Margarita Pacheco, 'Colombia's independent recyclers' union', in this book). Many industries, such as paper, cardboard and glass manufacturers, depend to a significant extent on recycled materials to reduce production costs. More contentiously, some even resort to importing waste. Chile, for example, imports containers-full of used disposable nappies from the USA to be recycled and used locally as industrial inputs.[20]

However with a handful of exceptions – such as the much publicised case of Curitiba in southern Brazil (see 'Curitiba: towards sustainable urban development' by Jonas Rabinovitch in this book) – recycling has not proved successful as a way of dealing with solid waste in Latin America's cities. For example, of the 10,700 tons of daily domestic waste collected in the city of São Paulo, a mere two tons is officially recycled.[21] Recycling is an important industry from an economic, social and environmental perspective, but its success depends on the very low incomes of those engaged in it. It also depends on the social acceptability of recycling, the awareness of urban dwellers of its potential benefits and an adequate legislation and regulation to promote recycling as an alternative. It will be some years before it is seen as a viable option in Latin America. Even in the richer nations, where most recycling involves packaging of household consumables, recycling is an expensive option, whose costs outstrip those of collecting unsorted rubbish.[22]

If the urban environmental challenges now facing Latin America appear vast, no doubt they would be even bigger if governments alone, with their less than impeccable management track record, were to tackle them using old methods. Fortunately the range of institutional changes now sweeping through the region and the increased openness and responsiveness of representative structures offer a glimmer of hope. As Eugenia Rodríguez has noted 'we have come to realise that, although environmental problems are a responsibility of governments, the attitude of every citizen has an impact upon the environment; this is a challenge for every one of us'.[23]

The urban environment in a changing institutional landscape

Cities and city governments stand at the core of the wide-ranging political changes that have swept through Latin America since the early 1980s. Elected national, provincial and local governments, effective decentralisation programmes and frequent exercises in public consultation and participation such as that of Lima's San Juan de Miraflores district (see Box 1, following page) are today the rule rather than the exception they were in the mid-1970s. In some ways, the urban environment has been both a catalyst and a beneficiary of these changes.

The environment was placed at the centre of urban political activity in the 1970s when unprecedented numbers of poor urban dwellers living in illegal subdivisions, squatter settlements or inner-city slums voiced their demands, especially for basic services like water and sanitation, but also for land, shelter, transport. The *barrio* became the privileged location of political activity, thus providing a springboard for budding politicians and community activists, as well as a fertile ground for the vigorous expansion of non-governmental organizations (NGOs). During these decades NGOs played an active part in channelling technical and organizational expertise and also international financial resources into the improvement of living conditions amongst some of the poorer urban communities. Even under repressive military governments, many also helped keep alive some form of local representative political activity and democratic participation, otherwise non-existent in national politics.

The political reforms of the 1980s gave renewed status to citizens' and communities' pressures; governments now have a greater capacity to react to these pressures (especially at the municipal level) and to interact with the population than in previous decades. With mayors and other local government bodies now the subject of periodic elections, the ballot box has become a powerful mechanism for interaction with voters and an incentive for greater local accountability. Representatives of stale political parties are shunned in favour of 'civic' mayors, who lack the backing of political *caciques* but offer an attractive break with the past. Elected officials regularly consult their citizens on matters of relevance. Unprecedented decentralisation programmes mean that municipal governments

BOX 1:
District planning with community participation in urban Peru

In the late 1980s some of the most innovative attempts at including local communities in the running of their cities took place in Peru. One of these was a development plan for San Juan de Miraflores, a fast-growing low-income settlement in the south of Lima with a population of some 300,000 in 1991. The area has a serious water supply problem, sewage and waste collection are either deficient or non-existent, and the municipality lacks adequate resources and plans to make real improvements. Like many poor settlements throughout Latin America, San Juan is a victim of unplanned expansion and an inadequate local tax base.

With the support of a £10,000 grant from Save the Children Fund, the Peruvian NGO, IPADEL,[24] completed a diagnostic study of the area. It also opened up a consultation process with the local population and local entrepreneurs to gather ideas and information. The main proposal to come out of this was that San Juan and two neighbouring districts be developed as a commercial and financial centre for southern Lima, with more detailed proposals regarding land use, transport and roads, services, housing, environmental health and employment.

Despite community backing, a lack of support from San Juan's new mayor (elected in 1989) stalled implementation of the plan, highlighting the difficulties posed by political factionalism in Peru. But the experience of drawing up a district plan based on the real aspirations and needs of the population was an important achievement. With a proliferation of community-based initiatives in Lima's squatter settlements, this project was an important step towards ensuring that communities influence local government policy making at a higher, district level.

Source: Elsa Dawson, 'District planning with community participation. Peru: the work of the Institute for Local Democracy', *Environment and Urbanization, Vol. 4, No. 2*, October 1992.

have more responsibilities, in theory at least, and can respond more adequately to local demands. There is a new political culture in Latin America's cities.

BOX 2:
Golf and the urban environment in Tepoztlán

In September 1995, protests from local residents, small businessmen and national and international environmental activists forced the resignation of the mayor of Tepotzlán and looked set to de-rail plans for a golf course and an industrial estate involving Mexican and US firms. An opinion poll showed 80 per cent of *tepoztecos* opposed the plans, amid allegations that the project was designed for wealthy outsiders and foreigners, and would create water shortages, increase living costs and change the character of the town. The project is located in a national park adjacent to the town, an hour's drive south of Mexico City. Upon presentation of an environmental impact assessment study in 1994, the National Environmental Institute (INE) had granted planning permission, subject to modifications. Protests erupted among a broad coalition of town dwellers when it emerged that these modifications had not been introduced and the worksite had been temporarily closed by the environmental authorities.

Source: Various issues of the Mexican daily *La Jornada*

The urban environment has benefited from these reforms in more ways than one. Firstly, communities have become aware that they can take a more active role in protecting their own environment, even in the face of national and international pressure, as the case of the Tepozteco Golf Club has shown (see Box 2). Secondly, representative local governments have understood that a concern with the urban environment is also a concern for the quality of life of its citizens rather than a passing fad. This was no doubt helped by the much-publicised launch of Agenda 21 at the Earth Summit in Rio de Janeiro in June 1992. Municipalities throughout Latin America are now engaged in supporting environmental initiatives such as research and monitoring of air quality, improved waste management schemes and even education campaigns for civil servants and the wider public (see Box 3). Several municipal governments are now preparing 'local environmental agendas' as suggested in the Earth Summit, often in consultation with resident communities, firms and NGOs and with the help of a growing number of expert groups based in universities and research centres. Bogotá's Environmental Department has produced detailed environmental plans for each of its twenty districts; implementation of the plan is overseen by a local environmental

BOX 3:
Environmental training for municipal officials in Quito

The Municipal Training Institute (ICAM) in Ecuador's capital city, Quito, has implemented a programme to raise environmental awareness among children attending state schools and poor communities and to provide environmental training for key municipal officials, including police officers. At the core of the programme is the belief that an involvement of social actors is a crucial step towards confronting environmental problems. In addition to producing a set of booklets, ICAM organized a series of workshops in four of Quito's poorest *barrios*, a play with teenage actors and a live radio debate. ICAM has received requests from other municipalities in Ecuador to develop similar programmes.

Source: Eugenia Rodríguez, 'Educación y capacitación ambiental en Quito', *Medio Ambiente y Urbanización, 1994 No. 47-48.*

commission comprising, among others, the district mayor, representatives of environmental groups and a high school student.

Thirdly, national and local governments, entrepreneurs and communities are slowly realising that a healthy urban environment is an essential step towards combatting poverty and improving productivity. This has prompted a strengthening of the tax base in many municipalities. Development plans in most of the larger cities now comprise an environmental dimension or an explicit environmental action plan, designed to raise awareness among ordinary citizens and industrial polluters alike, to improve waste management, and to monitor and control pollution. Mexico City's budget for pollution control virtually doubled between 1989 and 1994.[25]

Conclusion

Much remains to be done. In spite of decentralisation programmes devolving responsibilities from national to provincial and municipal governments and in spite of the increased resources available to medium and small-sized municipalities, the majority of these still lack the technical expertise and the money to monitor pollution, control deforestation, improve waste management and provide adequate basic services to their poorest citizens. Traditionally, local governments in Latin America have tried to confront environmental deterioration and pollution with yet more detailed master plans and tougher bye-laws. This approach has proved not only limited but also at times counterproductive: people simply shift their waste elsewhere, or dump it illegally; inspectors are bribed to turn a blind eye to unhealthy working environments; car users buy two cars with different number plates to by-pass official bans on traffic circulation. As governments are forced to cut costs, balance their books and reduce their workforce, their approach and capacity to directly influence the actions of others must change: the emphasis should now be placed on education and awareness-raising campaigns, on designing effective but realistic regulatory frameworks, rather than simply creating new and more detailed legislation that nobody will enforce. Governments should become more entrepreneurial and act as catalysts, coordinators and promoters of sound environmental initiatives.

[1] Economic figures from Friedrich Kahnert 'Improving urban employment and labour productivity', *World Bank Discussion Paper No. 10*, Washington, DC 1989. Population figures from J. Dávila et al., *Población y cambio urbano en América Latina y el Caribe, 1850-1989*, International Institute for Environment and Development, London and Buenos Aires.

[2] Most population figures in this section are drawn from national population censuses compiled in J. Dávila et al., op. cit.

[3] Ian Thomson, 'Los metros de América Latina', *La Era Urbana*, Vol. 2, No. 2., 1993.

[4] The contiguous built-up urban area of Mexico City increased from approximately 650 km^2 in 1970 to 1,115 km^2 in 1980, a growth of nearly 72 per cent in ten years (Boris Graizbord and Héctor Salazar Sánchez, 'Expansión física de la ciudad de México', in Gustavo Garza (ed.), *Atlas de la Ciudad de México*, Departamento del Distrito Federal and El Colegio de México, 1987.

[5] World Bank, 1992, *World Development Report 1992*, Washington DC.

[6] UNECLAC, 1994, 'Balance preliminar de la economía de América Latina y El Caribe', *Notas sobre Economía y el Desarrollo*, No. 556/557.

[7] The metropolitan region of Buenos Aires ('Polo Metropolitano') comprises the Federal Capital and 25 surrounding municipalities within the built-up area (cf. Francisco Gatto, Graciela Gutman and Gabriel Yoguel, 1987, *Reestructuración industrial en Argentina y sus efectos regionales, 1973-1984*, CEPAL, Paper No. 14a-CFI).

[8] Escobar Latapí, Agustín and Mercedes González de la Rocha, 1995, 'Crisis, restructuring and urban poverty in Mexico', *Environment and Urbanization, Vol. 7, No. 1* (April).

[9] World Bank, 1992, *World Development Report 1992*, Washington, DC.

[10] As Alan Gilbert has noted: '... how did (the administration of Mexico City) manage to accommodate five million additional people during the 1970s? How would the authorities in London, Paris or New York have managed if five million more people had been added to their cities' population within a decade?' (Cf. Gilbert, Alan, 1994, *The Latin American City*, Latin America Bureau, London, p.103).

[11] Biochemical oxygen demand (BOD) is a widely used measure of water pollution, as it provides an estimate of the amount of oxygen bacteria and other micro-organisms in the water needed to break down the waste.

[12] Augusto Pescuma and María Elena Guaresti, 1991, 'Gran Buenos Aires: Contaminación y saneamiento', *Medio Ambiente y Urbanización, No. 37.*
[13] World Health Organization and United Nations Environment Programme, 1992, *Urban Air Pollution in Megacities of the World*, Blackwell, Oxford.
[14] R Mejía Gastón, 1994, 'Manejo ambiental urbano: El Alto, Bolivia', *Medio Ambiente y Urbanización Nos. 47-48*
[15] Glewwe and Hall, quoted in Gilbert, 1992, op.cit.
[16] Jorge Hardoy and David Satterthwaite, 'Medio ambiente urbano y condiciones de vida en América Latina', *Medio Ambiente y Urbanización No. 36*, 1991. For a discussion of the links between a poor urban environment and health, see Jorge Hardoy, Diana Mitlin and David Satterthwaite, *Environmental Problems in Third World Cities*, Earthscan, London, 1992.
[17] United Nations Centre for Human Settlements (Habitat), 1996, *An Urbanizing World: Global Report on Human Settlements*, Oxford University Press, New York.
[18] Jiménez Díaz, Virginia, 1992, 'Landslides in the squatter settlements of Caracas; towards a better understanding of causative factors', *Environment and Urbanization Vol. 4, No. 2.*
[19] María Di Pace, Sergio Federovisky, Jorge E Hardoy and Sergio Mazzucchelli, 1992, *Medio ambiente urbano en la Argentina*, Centro Editor de América Latina, Buenos Aires.
[20] Sofia Torey, 'Reciclaje en Chile (I): La urgencia de un rayado de cancha', *Ambiente y Desarrollo Vol. XI, No. 2* June, 1995.
[21] Pedro Jacobi, 'Households and environment in the city of São Paulo; problems, perceptions and solutions', *Environment and Urbanization Vol. 6, No 2, 1994.*
[22] It is estimated that recycling half the reusable rubbish collected from Britain's household bins could cost twice as much as ordinary collection. Cf. Frances Cairncross, 'A survey of waste and the environment', *The Economist*, 29 May, 1993.
[23] Eugenia Rodríguez, 1994, 'Educación y capacitación ambiental en Quito', *Medio Ambiente y Urbanización Nos 47-48.*
[24] IPADEL was founded by former local government officials in 1987 with the support of five well-established Peruvian development NGOs (DESCO and CIDAP among them). It acts as a 'think-tank' and actively encourages local democracy.
[25] Javier Beristain, 'New strategies in financing city development: Recent experiences and challenges in managing Mexico City's finances', paper presented at the conference *Management of Cities during Structural Adjustment*, Bombay, 12-14 October, 1995.

San Salvador: The city versus the forest

Nick Caistor

San Salvador has all the ugly features of most Latin American capital cities. Greed, lack of planning, natural disasters and civil violence have all contributed to its cramped, unattractive appearance. The old downtown area was largely destroyed in the 1986 earthquake, and much of the centre has been taken over by needy people displaced from the countryside by almost two decades of conflict. Both these factors have helped accelerate the growth of the city up a hill towards the north and west. The middle classes rarely venture outside the 'better' areas of Escalón, San Benito or the Zona Rosa. These neighbourhoods are now full of new shopping centres, cinema complexes and hamburger restaurants where those with money can eat under the watchful gaze of shotgun-toting security guards. The suburbs are swiftly eating up the distance between San Salvador and what used to be the completely separate town of Santa Tecla, situated on the slopes of the San Salvador volcano, on the Panamerican Highway out towards the Pacific coast at La Libertad or Acajutla.

It is on this road out of San Salvador that the El Espino *finca*, or coffee plantation, is situated. El Espino (meaning 'The Thorn Tree') is almost 1,800 acres of secondary woods benefiting from the high terrain (over 400 metres) and the rich earth on the slopes of the San Salvador volcano. Although not primary growth, the coffee plantation offers a habitat for tall trees which serve as shade for the coffee bushes. The regular care and attention paid to the cultivation of coffee thus helps guarantee the continuing health of the forest. The finca is also home to a rich variety of bird and animal life, in particular to the parrots which have given it the nickname, 'Los Pericos'. The history of El Espino is a violent one, and the dramatic changes it has seen over the past twenty years show how closely ecological concerns intertwine with political and social factors in Latin American society. The pressure on El Espino is also a stark illustration of the financial speculation to which property close to Latin American cities is now prone. Land can be as explosive an issue for expanding urban populations as it is for their rural counterparts.

The Espino finca belonged to the Dueñas family. The Dueñas are one of the traditional fourteen families who own land, businesses and properties to the detriment of almost everyone else in El Salvador. According to one source,[1] in El Salvador in the 1970s, 1.5 per cent of holdings took up 49.5 per cent of available agricultural land. According to the same author, in the same period 21.8 per cent of the active rural population in El

Salvador had no access to land at all, while a further 28 per cent had less than one hectare (2.47 acres) to exploit. Violence was at the root of the ownership of El Espino: it is said that the Dueñas family took over the property after arranging the killing of the previous owner, General Gerardo Barios, in the last century. For many years, members of the Dueñas family ran the plantation as a profitable coffee farm for export. They employed a regular workforce of several hundred tied peasants, and also employed several thousand casual labourers at harvest time. Many of these labourers settled on the land, paying rent and growing what subsistence crops they could.

It was in the early 1970s that violence returned to El Espino. Known as the 'Tom Thumb' of Central America, El Salvador is a small, densely populated country. Because of the increasing use of agricultural land for export crops such as coffee or cotton, the rural poor found it increasingly difficult to support themselves from the land they were left with. As they organized to press for their demands, so the landowners resisted, with escalating violence. One of the Dueñas family who owned the finca was killed, and many of the others chose to leave El Salvador and set themselves up in the safer haunts of Miami. From the mid 1970s, although still run by the family, none of them was present to direct work on the finca.

As the violence increased in the late 1970s, the United States became increasingly drawn into the conflict. Successive administrations backed Salvadorean military and rightwing governments with many millions of dollars in order to prevent a takeover by leftwing guerrilla fighters. They also applied tactics used in Vietnam, and pushed a land reform on to the Salvadorean authorities. This land reform was to be in three stages. First, the expropriation of all properties over 500 hectares. This was carried out in March 1980.[2] Out of a total of 472 properties affected, comprising some 15 per cent of El Salvador's farmland,[3] fourteen belonged to the Dueñas family. One of them was the El Espino finca.

Ownership of the finca was handed over to a cooperative of two hundred or more of the regular workers there. Ricardo Orellano, president of the cooperative recalls: 'They told us that this was now our land and that we should defend it. That is what we have been doing, but there has been a lengthy campaign to evict us and give it back to the old owners.'[4] As well as this legal battle for ownership of the estate, the cooperative had to face other challenges. In a move typical of the time, almost one hundred acres were taken over by the army. They set up a barracks on the land, both to watch over the main road out of the capital, and to keep an eye on the activity of the peasant farmers, whom they regarded as potential guerrilla sympathisers. This was not a propitious moment for launching the new cooperative venture. The increasingly violent civil war disrupted coffee production; there was no credit available for the farmers; and although the new owners knew the finca well, they had little expertise in running the business.

Nor did the Dueñas family accept the expropriation. They continued to fight the decision legally. In 1987 the El Salvador Supreme Court reversed its orginal decision. The judges now ruled that the finca was part of the urban territory of the capital, and as such should have not been included in the original land reform. José Francisco Méndez, a lawyer employed by the cooperative, claimed to one British journalist that the decision was the result of bribes: 'We have talked with people who have assured us that a large amount of money was handed out. People talk about two million dollars. The decision was made on the testimony of two ex-government officials who later started working as surveyors for the Dueñas family.' [5] The Supreme Court judges and the Dueñas family have always denied these allegations. Because of the continuing violence, there were no attempts to enforce the legal decision immediately.

The situation was further complicated in 1988, when an environmental protection order was passed, covering the El Espino estate and the nearby slopes of the San Salvador volcano. The order reflected the government's belated alarm at the extent of environmental damage in El Salvador. Years of civil war had led to a vast amount of the country's forest cover either being burnt or bombed in the army's attempts to flush out guerrilla forces. At the time, concerns for the environment and wildlife had been the lowest priorities for both government and guerrilla forces alike, with the result that El Salvador's once rich variety of plant and animal life has been largely destroyed. By 1988 environmentalists were also arguing that the El Espino cooperative was of vital importance to the inhabitants of San Salvador. The forest maintains the level of the water table for the region and is thus crucial to the supply of fresh water to the city (the Lempa river close to San Salvador is already seriously polluted with sewage and chemical wastes). The tree coverage in itself produces rainfall and stabilises the weather conditions in the immediate vicinity.

It was only after 1989, when the right-wing ARENA party president Alfredo Cristiani came to power, that there was renewed pressure to resolve the situation at El Espino. The protection order was lifted. Cristiani's agriculture minister Antonio Cabrales justified the decision to repossess the estate in the following terms: 'When you have a decision by the Supreme Court, the only thing you can do is obey it. If you don't, then you are not creating a law-abiding type of atmosphere. And President Cristiani's government is precisely that. From the first day we took office, we said we wanted to restore law and order.'

The deal the government was now offering was a compromise, but one which greatly favoured the Dueñas family. They were to be given back 20 per cent of the original area, on which they would be allowed to chop down the trees and build housing. The government was to pay them US$12 million for the remaining 80 per cent of the estate (although the family had set the value of the whole property at only US$2.5 million for tax purposes in 1989). A new element was introduced: the government proposed that part of the land it was to buy would be kept as a recreational area for use by citizens of San Salvador. Cynically, the authorities claimed that this was more beneficial to the overcrowded city than a coffee plantation. The rest was to be retained by the cooperative, who were also given the right to build on a small part of it.

This decision split the farmers. Some of them, supported by one of the factions of the FMLN guerrilla forces, argued that they should accept the deal. The rest of the cooperative, now numbering only about 170 associates, thought they should fight the entire process. They claimed the government was moving them to a part of the estate without running water or electricity, both of which were needed to process the coffee.[6] They called for an independent environmental study to be carried out, maintaining that to build on any of the El Espino land would only increase the country's already horrific environmental problems.

The dispute over the finca dragged on through the early 1990s. On one side were the ARENA government, and the Dueñas family. On the other, the members of the cooperative and the ecological organizations which sprang up as peace returned to El Salvador. These organizations were formed by urban activists, many of whom had been forced into exile during the worst of the fighting in the 1980s. They returned to El Salvador with experience of ecological groups in other countries, especially in the United States, and were keen to continue this work in their own country. Finance for environmental organizations such as CESTA (headed by the charismatic, environmentalist, Ricardo Navarro) has come from international sources, which has led to repeated charges by the Salvadorean authorities that these 'experts' were merely using local issues to fight political battles. El Espino has been one such case. Over the past few years, the environmentalists have made the issue one of national importance, with debates on the radio, in the press, and even in the national parliament. They joined with cooperative members to take direct action to stop any building. They organized mass petitions, and used legal mechanisms to delay the start of any work.

Then finally, towards the end of November 1995, the issue was apparently resolved. The Dueñas family, the Salvadorean Agrarian Reform Institute, and the cooperative signed a comprehensive deal. Under its terms, the Dueñas family will keep some 350 acres, on which they can build a new housing development. The government and the San Salvador authorities are to have over 200 acres, where they are to create the recreational park, to be called Los Pericos after the many parrots which nest in the trees. The El Espino cooperative is to keep some 1,200 acres, to be declared a forest reserve – apart from almost 40 acres where cooperative associates now living on land ceded to the Dueñas will be resettled. The cooperative can also either build on their land or sell it for development so that they can generate capital to pay the government for the rest of the land they are allowed to keep. Environmental activists have denounced the deal, saying it will pave the way to further disaster. They claim that members of the cooperative were blackmailed into accepting this solution: the cooperative itself insists that it was the best they could achieve after all these years of dispute.[7]

The new 'solution' to the problem of the El Espino estate is typical of the new El Salvador being constructed after the formal signing of a peace treaty in 1992. The original landowners have been bought off. They stand to make much more money than they could have hoped for if they had simply continued to exploit the estate as a coffee farm. Faced with the acute problem of the rapidly expanding population in the capital city, the government has given the go-ahead not for the cheap kind of housing which is most needed, but for luxury development which can only serve to increase still further the divisions in society which have already led to so much violence. As a sop, they have offered a recreational park. Like the parrots, the other wildlife and the trees, the welfare of the coffee producers is very low on the list of priorities. Cooperative members are convinced that once part of the estate is built on, and new roads and facilities are constructed to help access, it will only be a matter of time before the government decides that the rest of the estate is urban rather than agricultural land, and sells that for development as well. In El Salvador, as in many other countries, money and power still outweigh environmental concerns.

[1] Charles D. Brockett, *Land, Power, and Poverty,* Unwin Hyman 1990; pp 67-83.
[2] For more details of this land reform project, see Jenny Pearce *Under the Eagle*, Latin America Bureau London 1981; pps. 231-235.
[3] Charles D. Brockett, op cit pp 156-157.
[4] Ricardo Orellano, quoted from unpublished interviews with Helen Collinson in May 1993 and Nick Caistor in 1995.
[5] Tom Gibb. Unpublished material.
[6] Helen Collinson, 'A Tale of Two Forests: Politics and the Environment in El Salvador', *Central America Report* (UK), Spring 1993, p 7.
[7] *La Prensa Gráfica*, San Salvador 29 November 1995, p 6-A.

Colombia's independent recyclers' union: a model for urban waste management*

Margarita Pacheco Montes

In Colombia, recyclers collect and manage urban waste independently. Men, women, children, dogs and horses work all through the night along the main streets of Bogotá and other large cities, collecting and selecting waste from the pavements, rubbish dumps and industrial and commercial areas. Over the past fifty years, one per cent of the country's population has generated its income from urban recycling. Nationally it is estimated that 16,500 tons of waste are produced each day, which amounts to 5,000,000 tons of waste per annum. Over seventy per cent of this waste is organic while thirty per cent of it is made up of materials such as plastic, paper, metal, glass, cloth and toxic waste. It is on this thirty per cent that the recyclers concentrate their recycling activities. The waste is collected by cooperatives and its sale is negotiated through a chain of intermediaries and industries. Solid waste management is thus undertaken by a network of agents with very little participation from the state, municipal authorities or refuse companies. But the evermore complex chain of intermediaries associated with the collection and commercialization of collected materials has increased the demand for new models of organization amongst the recyclers.

Silvio Ruiz is a recycler in Manizales. In 1986 he and other recyclers learnt that a new rubbish dump was to be created in their city from which they were to be excluded. They were then informed that the existing tip – their habitual workplace – would be closed down. In response, they decided to form the Prosperidad (Prosperity) Cooperative to increase their bargaining power in the market. This later became a model of organization and an example to other recyclers in Cali, Pasto, Ibagué and other Colombian urban centres. Over the following four years, forty groups of recyclers were formed across Colombia. The process coincided with the initiation of the Social Foundation's Recycling and Environmental Programme[1] which promotes an integrated model of solid waste management based on community participation and provides support for groups of recyclers. Through this collective organization, recyclers began to see improvements in their work conditions, their income and their quality of life. So positive was their experience that in 1990 the First National Congress of Recyclers was held. This led to the formation of the National Association of Recyclers (ANR), the purpose of which is to coordinate their work with the State and with NGOs and to lobby for government policies of benefit to recyclers. Seventy cooperatives and factories participated in the Second National Congress in 1991. During the Third Congress in Medellín, Espólito Murillo, Vice-President of the ANR, called for an expansion in the organization's brief:

> We are calling together delegates from 78 affiliated groups from Colombia and representatives from sister groups in Ecuador, Peru, Brazil and Mexico, and in this way we are celebrating the First International Congress of Recyclers. We are going to build a model of alternative development hatched from these organizations.

The following year the ANR won the Colombia Habitat prize in Popayán, as an acknowledgement of its contributions towards environmental protection. In 1995 the ANR was selected as one of the fifteen successful experiences in Colombia at the 'Best Practices' meeting for HABITAT II. The formation of the ANR has thus greatly improved the image of recyclers and their own self-respect. There is now considerable social recognition of recyclers' contribution to environmental conservation. The rubbish they collect is no longer regarded as a product which should be hidden; on the contrary, it is a valuable resource. Recalling his experience on the rubbish dump at Manizales, Silvio Ruiz declares:

> We recyclers have contributed to the conservation of natural resources in Colombia by: gathering recyclable material; halting the felling of twenty trees for each ton of recycled paper collected; and conserving water and electricity. We reduce the volume of rubbish disposed of, increase the useful lifespan of rubbish tips and prevent environmental pollution.

Rodrigo Ramírez, director of one of the cooperatives, declared at the Second National Congress of Recyclers:

> We are the promoters of a dignified and honest model. We have nothing to be ashamed of. Being a recycler does not only mean earning a living through hard and tiring work, but also means rescuing the nation, by cleaning up street corners, rubbish tips and places where people hide the waste of excessive consumerism. We are rescuing the fauna, the air, the sources of water... rescuing life itself... Our situation does not differ much from one city to another... It does not differ because we work in an honest way. We have the right to a cleaner country, to a bed, a roof and clean bread and many of us do not have those rights. We don't have them because society has turned its back on us. In spite of the risk from infection, of wounds and

accidents, we continue to be recyclers, dedicated to building our organization so that our job may be respected and supported.

Nowadays, ANR members recycle approximately 1,100 tons per day, which constitutes five per cent of the solid waste produced by the country. According to the ANR's *El Reciclador* bulletin (The Recycler), in 1993 ANR members collected 400,000 tons of cardboard and paper, which meant eight million trees escaped the chop in that year (the rate of deforestation in Colombia is estimated at one million hectares per year). For each ton of recycled paper, the paper industry saved 2,000 litres of water which means a saving of 800 million litres per year. The Peldar report estimates that the recyclers provided sixty per cent of the country's glass between 1982 and 1990, saving the country nearly three million gallons of fuel.

The ANR: a model for urban development

At present, there are ninety groups affiliated to the ANR, 12% of which are in Bogotá, 4.4% in Medellín, 2.2% in Manizales, 24% on the Atlantic Coast and the rest in other cities in the coffee-producing region of Valle, Tolima and Huila. In all, nearly 50,000 Colombian families belong to the Association. With the support of the Social Foundation, the ANR has implemented a dynamic model of social development, based on community participation. Among the social projects supported by the ANR is the Integrated Centre for the Care of Recyclers (CAIR) in the district of La Candelaria in Bogotá, which houses a nursery, a children's school and a medical consultancy. Other similar centres operate in Cali and Barranquilla and each one is administered and part-funded by the local groups affiliated to the ANR. In CAIR in Bogotá, the parents pay a monthly quota equivalent to 800 pesos (one dollar), while the upkeep of the centre is supported by the Christian Children's Foundation; the district's Social Welfare Programme provides food for the children. The spacious and well-lit house in La Candelaria where CAIR is based was acquired through funding from the Social Foundation and Mobil Oil. As well as basic childminding, these centres give children education and health care relevant to their parents' occupation. One aim is to prevent the children's exposure to toxic materials.

The ANR has also initiated some experiments in urban, organic agriculture. Organic waste is decomposed into compost or humus and used as a substitute for chemical fertilizers in the cultivation of vegetables. This has not only improved the nutrition of recyclers' children but has also enabled recyclers to sell the organic produce on the commercial market. Such initiatives help diversify and increase recyclers' incomes. Many of the enterprises now being promoted by ANR-affiliated groups are quite ambitious. In 1993 the Nueva Generación (New Generation) and Nueva Esperanza (New Hope) cooperatives in Pasto on the altiplano of Nariño set up a factory which turns plastic waste products into pipes. In the same year *El Reciclador* reported that the El Porvenir (The Future) cooperative and sixteen affiliated members had been developing a system to manage hospital waste since 1991. The group began their work in the San Ignacio Hospital in Bogotá, then offered its services to the Misericordia Hospital and later to the Haveriana University, another Jesuit organization. In recognition of their success, they were asked to expand their work to the institution's laundry service. Meanwhile, the Northern Coast Regional Association (ARCON), representing eleven recycling groups in 1995, has successfully coordinated the sale of recycled glass to local industry, achieving an increase of 380 per cent per annum in the price of this material.

The example of Manizales deserves special attention. Here, recyclers in the Prosperidad (Prosperity) and Mejorar (Improvement) cooperatives (both affiliated to the ANR) have formed a commercial company called Ciudad Verde (Green City) which has managed to negotiate a contract with the city's public companies. Green City will set up and operate a recycling treatment plant on lands belonging to the municipality. It will also maintain the rubbish dump in Manizales, thereby increasing the dump's lifespan. The project has received US$150 million from the FOSES Fund of the Presidency of the Republic and US$390 million from the Social Foundation. An executive committee, with representatives from all areas, maintains control over all of the project's activities. At present, an investigation into the project is being undertaken, financed by the Dutch government through the Urban Waste Expertise Programme (UWEP) and coordinated by the Dutch organization, WASTE. In collaboration with the Institute of Environmental Studies in the National University of Colombia, the investigation aims to come up with more efficient mechanisms for managing the recycling plant and the 'Ciudad Verde' Company.

The mayor of Manizales and the recycling cooperatives are aware of the mutual benefits of the Ciudad Verde project. All concerned share the same objective of decreasing the volume of rubbish destined for final disposal and protecting the region's beautiful Andean countryside, dominated by the snowy Ruiz volcano. The hope is that the Manizales model will encourage other municipalities to implement similar models. In the municipality of Chiquinquirá, meetings between local recyclers and the ANR have led to the setting up of a recyling service which offers collection, sweeping and environmental education services to local citizens. This programme promotes the initial separation of waste within the household, thus encouraging the creation and use of compost. Clearly the training and advice offered to the ANR by the Social Foundation and by other organizations have increased recyclers' capacity to negotiate with small municipalities and industries in small towns such as Armenia, Soledad and Popayán. Recently, the Quibdó town council asked the ANR for 'hands-on' advice on how to encourage organizations in their own shanty towns.

The role of the state

However, the achievements of the independent recyclers are still not sufficiently recognised by the state, even if there is now a recognition of their potential use in environmental planning. In spite of the ANR's efforts, the recyclers continue to live in poor conditions with deficient housing and high levels of social and environmental deterioration. Recyclers are frequently exposed to toxic waste, for example, which is still not properly managed. At an economic level, the increasing productivity of recycling should now be linked to the policies and economic strategies of the raw materials' market. Prices of the materials collected by recyclers should be controlled in order to safeguard their incomes.

Another major problem is that a huge state plan to separate waste at its source has not yet been put into practice. If implemented, it would allow for the collection of material in a more ordered fashion. To date, most municipal authorities have hardly begun to consider ways of incorporating waste separation into urban planning, despite its high economic, social and environmental value. Thirty per cent of the 6,000 tons of waste produced each day in Bogotá consists of recyclable material which could be used by Colombian industry. However, 1,800 tons is being buried each day, representing a daily loss of US$48 million (the value of each ton is approximately US$22,000). By dealing with the situation in this way, the state is increasing expenditure but not investment in waste collection. Greater collaboration between the state and the recyclers could be a highly profitable business.

The achievements of ANR-affiliated groups have led recyclers to question the ways in which they relate to non-governmental organizations and to reflect on the incentives they are seeking from the state. Above all, the environmental management model developed by the ANR and the Social Foundation should be supported by the Ministry of the Environment, by regional corporations, and municipalities. There needs to be coordination between national, regional, and local plans, both public and private;

universities should also be encouraged to participate in this process. Without a doubt, the adaptation of this solid waste management model to other cultural contexts and ecosystems will promote the urban sustainability of Colombia.

Translated by Judith Escribano

*Sections of this essay are taken from Margarita Pacheco, 'Empowerment of Independent Recyclers in Colombia' in C.E. Taylor, E.Desai, K.E. Knutsson et al eds, 'Partnerships for Social Development', Future Generations, Washington 1995.

[1] *La Fundación Social* or Social Foundation is a social fund set up and run by Colombia's Catholic Jesuit organizations.

Santo Domingo: an alternative city plan

Jorge Cela[1]

Culture and the environment

Santo Domingo, the capital of the Dominican Republic, is split in two by the River Ozama. Next to one of the bridges that crosses this river, there is an old building, now remodelled as a rehabilitation centre for sex workers. It was originally built during the dictatorship of Rafael Trujillo (1930-61) as a garbage incinerator for the city. It was then abandoned for thirty years and occupied by 108 families who had nowhere else to live. They lived in the basement of this incinerator, their only air and light coming through the doorway of the small staircase. It was almost impossible to breathe if you were unaccustomed to it. On a floor of charred rubbish, the inhabitants constructed partitions out of cardboard, fixed up their precarious electrical installations, and cooked on wood or charcoal stoves. There was no plumbing and they had to share one water tap. It is difficult to imagine a more hostile environment. As a result of living in such close quarters, there was a lot of aggressive behaviour.

When these families approached the 'Ciudad Alternativa' (Alternative City organization)[2] in 1989, they had no experience of organized action. They had arrived through COPADEBA (Committee for the Defence of Shanty Town Rights), to whom they had turned in the face of an imminent eviction. At that time, we were living under the government of President Joaquín Balaguer whose period of rule was characterized by the construction of monumental buildings. He had just begun issuing evictions in order to reconstruct the city centre (nearly 50,000 families were to be evicted during the ten years of 1986-1995). The population of the old incinerator had already made efforts to be relocated in a housing project. The government's only response was to threaten them with eviction, offering them a payment of less than US$50 compensation.

We hesitated when asked to help these people. We had always helped groups to manage their environment by themselves. But few of the incinerator inhabitants had had any experience of organizing. Most had never even had a stable family environment, let alone a stable institutional one. We wondered whether a group of human beings could be affected by poverty and their living conditions to such an extent that it could prevent them from organizing and managing themselves. In spite of their lack of organization, the families came up with a proposal to secure state funding for a self-building project. To follow through such a proposal, they were going to have to enter into protracted negotiations with the state from a position of weakness. But if their efforts proved successful, it would set a precedent for state support for such projects. We decided to accept the challenge, conscious of the fact that action on the environment can change the group which inhabits it. The incinerator inhabitants had been transformed by their environment and, in turn, this transformation would prompt them to change their environment. The presence of an advisory institution and a popular organization could help provide them with the necessary knowledge about management which would carry their project forward.

Keen to transform their habitat, those living in the incinerator began to hold meetings: they agreed points for negotiation and elected delegates to represent them. As a result, the group managed to formulate a proposal for the relevant government office (Bienes Nacionales/National Welfare) and to negotiate an agreement which provided them with plots of land and granted them materials to build their houses. With the help of Ciudad Alternativa, plus joint finance from international agencies and from the state, the inhabitants built their own homes. The process required patience and steadfastness in the face of bureaucracy and official lack of interest. The local population had to learn new skills, not only in building but also in negotiating. After years in the filth and darkness of the incinerator, they now felt as though they were owners of their own homes and began to improve them, to paint them up and to tidy up their gardens. By using the confrontational skills they had acquired from negotiating with the state, they began to reclaim basic services such as electricity and streets.

Once they had finished the project, they returned to Ciudad Alternativa to apply for advice on draining a gully for sanitation purposes and to construct a community centre which might serve as a nursery. The children who had been born in the incinerator, many of whom were now teenagers, formed a sports club and built a basketball court. Later they went to an open meeting in the Town Hall to ask for sports equipment which they have now received. This is just one example of a community's conversion into a group capable of managing the improvement of their environment in consultation with the state and with specialist service institutions.

Two main lessons can be learnt from this experience: the environment changes people's behaviour and encourages new forms of social coexistence. The participatory management of the environment is not only an efficient and permanent solution, but also acts as a training school for the citizenry.

Barbarism, civilization and paradise

We have frequently described history as the bitter struggle between barbarism and civilization. Of course, the barbarians were always 'the others'. Devoid of rational interpretation and behaviour, they were responsible for backwardness, violence and destruction. Civilization, linked to the rationality of scientific knowledge and educated behaviour, represented peace and progress for all people. In discussions on the environment, we often use these manichean terms, believing barbarians to be the predators. They are the ones who are responsible for deforestation, for contaminating rivers and for producing the waste which pollutes the city. The poor are the new barbarians, the scapegoats who carry all the ecological sins of the world on their shoulders.[5] Supposedly, their actions should be controlled through laws and repression and they should be expelled from reforested zones, riverbanks, oceans and all lands of any value.

The experience of the La Ciénaga shanty town in Santo Domingo is an expression of such a vision. This area has been deemed uninhabitable many times. Situated on the edge of the River Ozama, right in the heart of the city, and visible from the busiest bridge, the district has been condemned by President Balaguer as an embarrassment to the country. He holds the inhabitants responsible for the serious pollution of the river. From his perspective, environmental management means repressing the 'barbarians' and protecting Nature from its predator-inhabitants. This breaks the link between human beings and the environment and claims the environment as the absolute to which humanity should be subordinate. From this point of view, environmental management is seen to be the defence of Nature against human beings. In order to justify this concept, norms are created in which humanity is distanced from Nature. The state becomes the guardian of this distancing process.

The solution proposed by the state for the La Ciénaga district was to remove all of its inhabitants and militarize the area in order to prevent any building or repair work on houses or any other physical infrastructure.[6] This measure affected more than 13,000 families, and still remains intact, impeding the improvement of the district. The matter is so serious that many craftsmen in the district have gone bankrupt since building materials such as wood and cement are not allowed into the area.

Technical development has often been seen as the process by which human beings 'dominated' the environment. This is the argument which justifies the 'developed' countries' right to decide for others and the right of elites to decide for the majority. Meanwhile the poor are excluded from participation in the elaboration of policies which affect them. The exaltation of technical knowledge is linked to the myth that continuous progress shaped the modern world. In reality, huge imbalances and ruptures were caused, including the breakdown of the harmonious relationship with Nature.[7]

With this concept of modernization in mind, a park surrounding an avenue was designed by government officials to replace the district of La Ciénaga. They used environmental arguments: they claimed they were rescuing the highly polluted river from the people living on its banks who were supposedly the source of the pollution. But they also used modernization and developmental arguments: the project would improve traffic flows in the city. It would also be an investment for the state to support projects which would encourage tourism. By removing the poor from the city centre to the outskirts, they would be no less poor but they would be less visible.[8]

Environmental management should be oriented principally towards the improvement of the quality of human life and not vice versa. We should not place human beings at the service of the environment, but the environment at the service of human beings. Environmental management should never conflict with a determined group of human beings. But at the same time, an improvement in the quality of human life cannot be achieved without a healthy environment and harmonious integration within it. If we accept that human beings are the subject of such a relationship, then they must be granted a degree of participation in the management of the environment. This management is part of what Lechner has called 'the interior courtyards of democracy'[9]: the multitude of relationships which shape the true participation of the citizenry in the life of modern society. Ultimately, it shows us that management cannot exclude or discriminate against any social group.

This is how it was understood by some of the inhabitants. Organized groups within COPADEBA from the Ciénega district approached Ciudad Alternativa, requesting technical advice to resolve some of their problems. A dialogue was initiated with the inhabitants and a sketch of a design for improving the district was drawn up. The process began with listening to the inhabitants. They themselves identified the uninhabitable zones and the main problems: the contamination of drinking water, areas liable to flooding, the collection of rubbish, and insufficient drainage. They designed the first emergency solutions to the problems and they negotiated with the state: they built two sewerage systems and they mended a drainage channel for rain water. They carried out workshops with teachers in the school and they organized study and action groups with primary school children in order to increase their consciousness and to foment positive attitudes towards the care of the environment. An inhabitants' study group started working on the problem of the river. Data was compiled to demonstrate that the inhabitants were not the source of the river's problems. The city's sewage poured into the river without being treated, and an industrial zone with more than 150 industries threw its waste straight into the river.

As a result of this work, a team of professionals started to draw up a draft proposal for urban improvement. The inhabitants' study groups continued to meet so that they could feed ideas into the proposal: while negotiations ensued with the State, the local population was busy trying to establish a public consensus. As part of this process, they accepted an invitation to present their proposal to a large audience on a TV programme. The response from the state did not give us much hope. The Legal Consultant of the Executive interrupted the broadcast in order to read out a decree to evict the population. The attempt at participation was therefore ignored and substituted with an order which provided no solution for the 13,000 families inhabiting the area. Four years after the decree, only some 400 families (about three per cent of the district's population) have been evicted. But the militarization of the area has been maintained and this has impeded the involvement of the inhabitants in the search for solutions. Furthermore, the frustration of not securing the housing they had hoped for has created divisions amongst the population.

The most effective formula for technological advance does not create ruptures or an imbalance in the relationship between human beings and the environment. This formula consists of increasing democratic participation in environmental management so that everyone can contribute to the search for viable and effective solutions. Environmental management is a function of every citizen and not only of the state. Coordination between the state and civil society is imperative. There should be collaboration with institutions representing active social sectors, such as non-governmental organizations and academic institutions that can contribute scientific knowledge and other information to specialist state institutions. In the case of La Ciénaga, the Dominican government disregarded not only the affected population but also the work of numerous professionals from Ciudad Alternativa and from national universities and even some foreign universities.[10]

Recent research suggests that as the population becomes more informed and acknowledges its responsibility towards the environment, it also becomes more involved in its protection and more efficient in such involvement.[11] [12] This was not the outcome in La Ciénaga. The only work which the state has undertaken is the construction of a sewage treatment plant right in the middle of the district which has been built without reference to laws requiring the creation of a green area around such constructions. Once again the inhabitants have been marginalized, leading to protests and to violent rebellion, in the absence of any real channel for dialogue. The state has responded with repression which has already cost three lives. This has resulted in delays in the building of the plant and an increase in costs. The inhabitants have repeatedly destroyed new construction work on the plant and prevented the building from continuing. For months they have lived under the shadow of a possible internal war. Nothing has improved in the shanty town. Meanwhile, Ciudad Alternativa continues working on the elaboration of an alternative proposal which takes into account the protagonists of this drama: the inhabitants themselves.

How to manage the environment

Co-management of the environment has six dimensions which we should bear in mind: political, communicative, legal, educational, anthropological and economic.

The political dimension

The concept of participation in environmental management requires the political will to develop a more participatory democracy. We have to create joint organizations drawn from the state and from the citizenry which should not be mere executors, but should also have a decision-making capacity. The population should make decisions and implement them. They should collaborate with non-governmental organizations which could be contracted in order to undertake the technical side of projects. For this environmental co-management to happen, decentralisation of the state is important.[13] Ideally it would be run through municipal institutions. This would allow for greater versatility in environmental management towards local needs and healthy competition between municipalities which would act as an incentive to creativity. These joint municipal organisms (state-civil society) should be assured autonomy, resources (or the means to achieve them) and integration in national projects. These could be monitored closely in order to guarantee quality of management. It is important that civil participation be truly representative.

Communication

An indispensable condition for the effective participation of civil society is adequate information. Nobody knows the problems or the local resources available to resolve them better than the affected people themselves. However, these local organizations often lack the information that would permit them to articulate viable proposals. It is important that information on the global perspective – of which the local problem is a part – be made available. Such information would include hidden causes, technical elements which might allow innovation, and accessibility to resources.[14]

What is needed is a two-way flow of information. There is information that the local population possesses and which could be shared. Other information could be provided by the state (global data) or by specialists. It is in the exchange of such information that more appropriate, viable and effective solutions can be found. In order to create systems of communication, advantage should be taken of the local requests which already exist. And although the shaping or identification of these information networks may be quite an effort and time-consuming initially, it will prove beneficial in the long run.

The legal dimension

Environmental co-management needs to be institutionalized. We live in a culture of unlawfulness.[15] Living on the margins of the law appears to be a condition of survival. Take the 'unlawfulness' in production and informal business, the violation of environmental laws by big factories and by state institutions themselves; or the dream of a large number of Dominicans to enter Puerto Rico or the US illegally.

We need to make laws which are in the interests of those affected by them. Urban legislation cannot be the same for residential areas as it is for precarious shanty towns. Trying to apply the same legislation merely promotes ignorance and places the most impoverished in the difficult situation of being 'illegal', that is to say, without rights.[16] Environmental management should be transferred to the institutions within civil society which are best equipped to deal with it, either because of their knowledge and professional experience or because of their degree of involvement in the matter. This institutionalization should thus depoliticize environmental management and include supervision on compliance with the law.

The educational dimension

Management of the environment implies a change in the population's attitude towards its relationship with the environment. Promotion of the improved protection of the environment should not only be aimed at individuals, but also at creating mechanisms and developing collective behaviour in order to protect the environment. This means that ecological education should be planned in such a way that it reaches every single person. Educational programmes should not be limited to the school curriculum but should also be designed for use in informal education and for transmission by the mass media. It should not consist merely of education on ecology, but also education on participation. It is worth noting that work has already been developed on this front by many non-governmental organizations.[17]

The anthropological dimension

The aim of environmental education is to transform the relationship between human beings and the environment. It is therefore a cultural process which seeks to overcome certain cultural perceptions. The first of these is the separation between human beings and the environment.

Just as working the land tends to create an intimate relationship between human beings and the environment, the city tends to distance human beings from the environment. When the environment is presented as being hostile, this distance is accentuated. The filthy or dangerous city marginalizes its inhabitants and confines them to small areas in which they can find security and a degree of well-being. Huge modern cities give the impression of being cold, insecure and distant. This alien environment does not encourage people to take care of it, since people do not appear to belong. This perception of a hostile environment increases when there is overcrowding and results in aggressive attitudes towards an anonymous and distant environment. Personal spaces which we take care of become ever smaller, and we restrict that care to maybe just the garden, to our own bedrooms, or to our own bodies.

Taking care of collective space requires the existence of a collective identity and territorial consciousness. The development of this collective identity requires cultural work on raising awareness of one's roots. Generally, this is achieved through the development of an organizational network which should be as broad as possible. The creation of such a network and awareness about environmental management will result in greater participation, creativity and efficiency.

The economic dimension

Environmental management requires economic resources and the costs are extremely high.[18] The resources can come from a variety of sources: from the local community itself, from the state, or from national or foreign agencies. The ideal situation would be a combination of these resources since this would increase the project's capacity and autonomy. The local community could contribute to the process with cash or by working. Often, a contribution in the form of labour, work tools or the collection of funds can be significant. For the most part, we cannot talk of environmental co-management if there is not some economic contribution from the local population. However, neither can we talk of co-management if this is the only form of participation. The state should at least channel some of its funds destined for the environment through the local community, thereby strengthening its potential. Communities need to be informed about what state resources are available for them and to be aware of which channels are open to them. Strict control and auditing of the use of these resources should be carried out. Administrative decentralisation should not be the decentralisation of corruption. However, a community should not rely on the presence of these funds. They will not always be available for all communities and environmental management is a permanent necessity. This aid should be used only for extraordinary projects and not for the constant costs of environmental management.

We must promote pilot projects that will demonstrate the value of these proposals. There is also a need for negotiation between community organizations, non-governmental organizations, state agencies and aid agencies. The experience of Ciudad Alternativa confirms the validity of this approach because 'the city belongs to us all'.

Translated by Judith Escribano

[1] See also Jorge Cela, *Por una Ciudad con Participación Popular*, Ciudad Alternativa, Santo Domingo, 1991, and 'La Conquista del Espacio' in *Estudios Sociales*, no.88, 1992.
[2] A non-governmental organization dedicated to urban improvement through popular participation, created by the shanty town organization, COPADEBA (Committee for the Defence of Shanty Town Rights).
[3] 'Big factories and public works wear away, break down and destroy nature on a huge scale,' declares Emilio Pradilla Cobros, in 'Acumulación de Capital y Estructura Territorial en América Latina' in M. Lungo's *Lo Urbano*, Editorial Universitaria Centroamericana, Costa Rica, 1989, p.46.
[4] See V. Sánchez, J. Hurtubia, H. Sejenovich and F. Szekely, in *Hacia una Conceptualización del Ecodesarrollo*, p.28.
[5] For this concept applied to the Dominican situation, see Pedro J. Rosario and H. Schogmayer, *Medio Ambiente y Sociedad en el Espacio Rural Dominicano*, CEUR, PUCMM, Santiago, 1989, p.1.
[6] Decree 358-91 of 17 September 1991.
[7] For an example of the imbalance caused by the development of agro-industry, see Amparo Chantada, *Medio Ambiente*, MSC, Santo Domingo, 1992.
[8] The decree considered that only those families who were registered would receive housing. This meant that the poorest families would be excluded from the right to shelter.
[9] Norbert Lechner, *Los Patios Interiores de la Democracia*, FLACSO.
[10] A workshop-seminar entitled 'Un Mejoramiento Posible' was organized and attended by various architectural colleges from national universities and those from the University of Puerto Rico, the National University of Colombia and the Polytechnic of Milan.
[11] See V. Sánchez in M. Lungo, op cit p.32.
[12] See Christian Aedo and Osvaldo Larrañaga, *Sistema de Entrega de Servicios Sociales*, Postgraduate Economics Programme, ILADES Georgetown.
[13] See J. Borja in M. Lungo, op cit p.258 and Jorge Cela and Tahira Vargas, *Papel de la Sociedad Civil en la Reforma Social y Disminución de la Pobreza*, Santo Domingo, 1994, p.28.
[14] On the importance of information, see J. Borja, op cit p.28.
[15] Ordinary people are virtually barred from using the complex legal system because of the financial costs but also the costs in time and personal dignity. See Nora Clichevsky, 'Ciudad y Tierra Urbana' in M. Lungo, op cit, p.150.
[16] See Jorge Hardoy, 'Reflexiones sobre la Ciudad Latinoamericana' in M. Lungo op cit p.328. Interesting studies have also been carried out in the Dominican Republic by the Foro Urbano and CEDAIL-Ciudad Alternativa on the subject.
[17] See, for example, the United Nations Development Programme, *Informe sobre Desarrollo Humano 1993*, CIDEAL, Madrid, 1993, p.101.
[18] See Jorge Hardoy in M. Lungo, op cit pp.85-86.

Curitiba: towards sustainable urban development*

Jonas Rabinovitch

Curitiba is the capital of Paraná, a mainly agricultural state in the south of Brazil. Located 900 metres above sea-level, near the coastal mountain ridge, the city has long been an administrative and political centre. It developed rapidly in the second half of the nineteenth century, underpinned by the arrival of many (mostly European) immigrants, the expansion of urban services and the opening of new economic frontiers; Curitiba was a central location for roads and railways and became a key service centre for new economic activities. It was designated the state capital in 1854. Rapid economic and demographic growth in recent decades has transformed the city into an important industrial and commercial centre. It has grown from 500,000 inhabitants in 1965 to 1.6 million today and covers an area of 431 square kilometres. Service and commercial activities account for about eighty per cent of employment. Average per capita income is approximately US$ 2,500 per annum. Curitiba continues to be one of the fastest growing cities in Brazil. Despite this rapid growth, substantial improvements in the quality of life have been achieved – for example, the innovative public transport system, the preservation of the city's cultural heritage, the large expansion in the number and area of parks and green areas, the integration of social programmes and environmental education, the innovative 'garbage that is not garbage' solid waste management system and the 'garbage exchange' programme.

The planning process

During the peak of Brazil's rapid urbanization in the 1960s, a decision was taken in Curitiba to concentrate on a planning framework which emphasized the integration of all the elements within the urban system and which centred on a transport system that gave primacy to meeting the population's transport needs – rather than primacy to those owning or using private automobiles (at this time, most Brazilian cities were being planned for cars and individual modes of transport). Initiatives undertaken by the local government in Curitiba allowed the city to plan, direct and control its growth process; it avoided large-scale and expensive projects but included hundreds of modest initiatives. In spatial terms, the key concept was to encourage Curitiba's physical expansion along linear axes, each with a central road and a lane for express buses. The aim was to reduce the concentration of employment in Curitiba's traditional city centre in order to return this central area to the pedestrian and preserve the city's cultural heritage. Restrictions were placed on new developments in the centre and the commercial and service sectors expanded outwards along the linear axes to the north, south, east and west.

The original plan for Curitiba came out of a public competition and the winning plan was made available to the municipal authority in 1965. In this same year, the municipal authority created the Curitiba Research and Urban Planning Institute (IPPUC) which was allotted the following functions: to draw up the master plan, develop studies and projects for the integrated development planning of the Curitiba metropolitan region, create conditions for the implementation, continuity and flexibility of proposals and coordinate local planning with policies at a regional, state and national level. One of the key actors in the success of Curitiba is Jaime Lerner who has served three terms as mayor of the city over the past 25 years and is now the Governor of the State of Paraná. Prior to his first appointment as mayor, Jaime Lerner worked at the Curitiba Research and Urban Planning Institute and during his first administration, the IPPUC's powers and responsibilities were considerably enlarged. During the 1970s, three important integrated elements influenced Curitiba's development: the rationalization of transport, the development of the road network system, and land use legislation. These initiatives were complemented by the development of the industrial city (a specially designated industrial area on the west of the city), which generates one-fifth of all jobs in the metropolitan area without causing any industrial pollution or significant environmental problems.

What makes Curitiba unusual is not so much that it had a coherent plan but that it was *implemented* and that this plan was integrated with an effective public transport system and with various other initiatives to improve the quality of life in the city.

The bus-centred public transport system

Each of Curitiba's five main axes has been designed with a 'trinary' road system. The central road has two exclusive bus lanes in the centre for express buses flanked by two local roads. Each side of this central road, one block away, are high capacity free-flowing one way roads, one for traffic flowing into the city, the other for traffic flowing out of the city. Another important element of Curitiba's road network is the hierarchy of roads. Each road is assigned a function in relation to its location and importance. There are the 'structural' roads along the five axes described above and 'priority' links which connect traffic to the structural roads. 'Collector' streets have commercial activity along them with all forms of traffic, and the 'connector' streets linking the structural roads to the industrial city. These four types of road form the skeleton structure of Curitiba.

A key complementary activity to the road system was the municipal government's acquisition of land along or close to the new transport axes, prior to their construction. This enabled the government to initiate high density housing programmes close to the transport axes; in all, some 17,000 lower income families were located close to these. Residential and commercial areas are thus matched with the availability of public transport, taking the pressure off the city centre and returning it to pedestrians.

Land legislation further encouraged high density occupation and commerce in the areas adjacent to each axis. The whole of Curitiba is zoned according to the kind of use to which the land can be put and the density of development permitted. On the land sites located along the structural roads, legislation permits the construction of buildings with a total floor area of up to six times the plot size. Developments with floor space of up to four times the plot space are also permitted close to other roads well served by public transport. This has reduced the distance of land sites from public transport and lured new commercial and residential developments out of the city centre, along each main road. For the first time, a new mass transport idea was created whereby the bus routes and the use of land adjacent to the routes were more important than the bus itself.

Despite having 500,000 cars (more per capita then any other major Brazilian city), Curitiba does not have a traffic problem. When the present transport system was initiated in 1974 under the first term of Mayor Jaime Lerner, the city decided to continue restricting the city-wide transport system to buses. The use of 'express buses' using exclusive bus lanes is far cheaper than subways or light railways and represents a more practical and affordable solution to public transport for a Third World medium sized city. In addition to the central bus lanes along the city's main axes, new bus lines were created and expanded as the city grew. A series of circular inter-district bus routes developed, to complement the express busways. Buses are colour coded: the express buses are red, inter-district buses are green and the conventional (feeder) buses are yellow. In 1979 a standard 'social fare' was introduced for all bus users which greatly benefited the predominantly poorer bus users on the city's periphery and enabled shorter journeys to subsidize the longer ones.

One of the key concepts in the transport system is the ease with which people can transfer from local buses to the express buses and back to other local buses. There are large bus terminals at the end of each of the five express busways where people can transfer to inter-district or feeder buses. One single fare is valid for all buses. Along each express route, smaller bus terminals are located approximately every 1,400 metres and are equipped with newspaper stands, public telephones, post offices and small commercial facilities. Here passengers arrive on feeder buses and transfer to the express buses or inter-district buses.

The relative costs of different public transport options

Public transport option	Capital cost per kilometre
Underground metro system	US$90-100 million
Light railway system	US$20 million
Curitiba's direct route busway system (using boarding tubes)	US$0.2 million

The latest innovation is the introduction of the 'direct' express bus system where there are fewer stops and where passengers pay before boarding the buses in special raised tubular stations. These run along the one-way routes on each side of the structural axes' central road. These new stations with platforms at the same height as the bus floors cut boarding and deboarding times; a rapid bus system with these 'boarding tubes' can take twice as many passengers per hour and three times as many passengers per hour when compared to a conventional bus operating in a normal street. The boarding tubes also eliminate the need for a crew on the bus to collect fares, which frees up space for more passengers. In this way, just one of Curitiba's express buses does the work of many traditional ones (see Box).

Curitiba's public transport system is used by more than 1.3 million passengers each day, attracting nearly two-thirds of the population. Twenty eight per cent of direct route bus users previously travelled in their cars. This has helped secure savings of up to 25 per cent of fuel consumption city-wide. Curitiba's public transport system is directly responsible for the city having one of the lowest rates of ambient air pollution in Brazil. Another effect of Curitiba's transport policy is the savings for inhabitants in expenditure on transport; on average, residents spend only about ten per cent of their income on transport which is a relatively low proportion for Brazil.

Since 1979, the introduction of inter-district lines, a standard fare for the whole network and new integrated terminals have allowed for the operation of a 514 kilometre network. Automatic fare collection, articulated buses and traffic lights which give priority to buses (operated by the vehicles themselves) allow the optimization of the system's operation and the low operating costs. New initiatives are constantly being sought to improve the system. In recent years a 'bi-articulated' bus has been developed with a capacity of 270 passengers. These have five lateral doors for passenger entry or exit which help to reduce boarding and deboarding times, when linked to the new boarding tubes. Sixty-one of these buses are now operational.

The buses operating within this integrated transport system are privately owned by companies which receive a concession from the municipal government to operate specific routes. These companies have to abide by the city government guidelines and monitoring policies. The bus fares go to the municipal bus fund (managed by the mixed capital company, URBS) with the bus companies paid according to the number of kilometres their buses cover. Each bus company received a permit to operate particular routes with operational details and timetables defined by the municipal authorities. The whole public transport system operates without any direct financial subsidy.

The industrial city

Located seven miles from the city centre, Curitiba's industrial zone is well integrated within the urban structure and is equipped with services, infrastructure, schools, housing, green spaces and transport axes connecting it to the rest of the city. One reason why the city authorities have been able to develop this industrial city without a high cost is that they own the entire site. Part of the site was already in public ownership when they began to develop it in 1972 while the rest was expropriated, before the infrastructure was installed (and thus before the value of the land increased). Part of the land has been allocated to low income public housing. With over 400 non-polluting industries, this industrial area now generates one-fifth of all jobs in the metropolitan area. Any industry wishing to locate in this industrial city must conform to local environmental legislation. Industries located within the metropolitan region of Curitiba account for 31 per cent of the industry in the state of Paraná and generate over 100 million dollars annually.

Water, sanitation and garbage: managing and recycling solid and liquid wastes

Curitiba Metropolitan Area produces around 1,000 tons of garbage each day, of which three quarters is generated within the city and the rest from thirteen neighbouring municipalities. In 1990, Curitiba received an award from the United Nations Environment Programme for two successful waste management programmes. The first, launched by Jaime Lerner in 1989 is the 'Garbage that is not Garbage' recycling programme which encourages city residents to separate organic and inorganic garbage for recycling and collection. Once a week, a 'garbage that is not garbage' lorry collects the materials which households have sorted. Over seventy per cent of the community now participates in the programme and its success is largely due to a city-wide environmental education programme which highlights the benefits of recycling. In all, two-thirds of the city's recyclable garbage is recycled, more than a hundred tonnes daily. Since the beginning of the programme, Curitiba has recycled some 13,000 tonnes of garbage. Just its paper recycling saves the equivalent of 1,200 trees a day. Apart from the environmental benefits, this recycling programme has generated other positive side effects. The sale of the recyclable garbage, for example, is reinvested in local social programmes. The city authorities have also provided jobs in the main garbage separation plant to the homeless and to those recovering from alcoholism.

The second solid waste management programme, also launched by mayor Lerner, is the 'Purchase of Garbage' programme. This is run in the squatter settlements (*favelas*) of Curitiba; around ten per cent of Curitiba's population lives in favelas, most of them on the outskirts of the urbanized area. In most favelas, there was no service to collect household garbage, usually because the settlements lack the access roads which permit the garbage trucks to enter them (for instance many are located on the bottom of the valley, close to the river). The residents would simply dump their garbage in open air pits or vacant plots where flies and rats could breed and thus increase the risk of certain diseases. There was no basic knowledge of hygiene and the inhabitants were often undernourished. To help deal with these potential health problems, the city introduced a programme where favela residents could 'sell' their bags of garbage in return for bus tickets and for agricultural and dairy products. This programme has led to a considerable decrease in city litter and has helped to improve the quality of life of the urban poor. The cost to the city authorities arising from the provision of bus fares and food for garbage are similar to what they would have to pay a private company to collect the garbage. At the time of writing there were 22,000 families in 52 communities involved in this scheme. Preventing garbage dumping in rivers, forests and valley bottoms is not just an important step towards environmental preservation: infant mortality rates have decreased substantially in these poor areas while the reduction in diseases has meant a saving for many families in expenditure on medical care.

With regard to water and sanitation, a relatively high proportion of the city's population is served, compared to cities of a similar size elsewhere in the Third World. Some ninety per cent of the population has treated, piped water supplies and sixty per cent lives in housing units connected to a sewage system. Curitiba is also developing an innovative sewage treatment system which makes greatest possible use of a system of lagoons located close to the rivers into which the water will be discharged. The initial lagoons will be anaerobic (where micro-organisms break down the sewage in the absence of oxygen) followed by aerobic lagoons and finally discharge of treated water into the river. This treatment system has substantial cost advantages over conventional treatment systems, although more conventional sewage treatment plants will be needed in certain areas where residential densities (and thus sewage volumes per hectare) are highest. Various other measures are being used to reduce the pollution of the Iguazú river and its tributaries, including an open air canal running parallel to the river which will be a protection against flooding and also serve as a stabilization pond in which pollutants can break down before entering the river. A pedestrian walkway and cycleway are being developed on one of the banks of this canal.

Preserving architectural heritage, the expansion of parks and the protection of green areas

The integrated land-use planning and transportation system has prompted an enormous expansion in parks and green areas and has helped perserve the architecture and culture of the city centre. In the 1970s the centre underwent a major refurbishment process. Many streets became pedestrian areas while old buildings and historic sites were protected, public squares upgraded and shopping and commercial facilities developed.

Owners of buildings designated as of historic value are permitted to develop new uses in their buildings, but not to fundamentally change the facade and layout. To compensate for this restriction, the municipal authorities permit owners of historic buildings to sell the development potential they have lost for use on another site. One of the city's most popular shopping malls developed in what was previously a foundry. A former gunpowder arsenal has been converted into a theatre and an old glue factory into an arts centre. What had previously been the army headquarters has been converted into a cultural foundation and the city's oldest remaining house has become a documentation and publication centre. The old railway station has become a railway museum while a stone quarry has been converted into an open air theatre. In 1991 a 'twenty-four hour street' was created downtown by city authorities in partnership with local merchants. Businesses here stay open 24 hours a day, seven days a week, thereby helping to sustain commercial activities in the city centre.

The city has a well defined policy and strong commitment towards preserving its woods and parks. In the past twenty years, more than 1.5 million trees have been planted in the city. The ratio of open space to inhabitant has increased from 0.5 square metres to 52 square metres which means that Curitiba has one of the highest averages of green space per urban inhabitant anywhere in the world. The *Guarda Verde* (the 'green guard' – a municipal corporation) protects and maintains the green areas;

the guards also keep the public informed about environmental issues and are trained for first aid. There are programmes to encourage community responsibility for care and maintenance of the parks – for instance the Associations of Friends of the Park formed by volunteers, the Boy Scout Bicycle Watch to promote and protect the parks, and the use by local schools of parks to promote knowledge about them and about ecological principles. Another innovative feature of Curitiba's green spaces is their integration with flood control. The parks not only provide recreational and aesthetic value but many have artificial lakes which provide flood control for the entire city. Each park is equipped with information centres on the local environment and ecology. A 90-mile (145 km) bike path mostly through the urban parks is nearly complete. A recent addition is the new botanic gardens on the site of an abandoned garbage dump, covering 17.7 hectares. It includes some of the last remaining natural fauna and flora of the region. A museum is to be included along with a research facility to study local flora and the cultivation of native and exotic species, including species from lower coastal areas and other regions of the country. The city also provides a free bus service during the weekends on its 'pro-park' line. These are 'retired' city buses painted green which carry people from downtown to the city's numerous parks (Brazilian legislation demands that all buses in use on the roads be less than ten years old).

Social services and environmental education

A lack of education is a key reason for environmental destruction. By providing environmental education, the city hopes to improve the quality of life of low-income households, especially the children, and to teach them to be responsible for their actions.

In the past, no infrastructure existed to support any kind of day care for children and adolescents in the favela areas. Most children wandered around their settlement unsupervised while their parents were away at work. To address this problem, the Infant and Adolescent Units (with the acronym, PIA which means 'kid' in southern Brazil) were set up. In the favelas and other low-income areas, simple huts were built with wood-burning stoves for cooking and heating which serve as drop-in centres for local children. For every 300 children there are two employees, making this programme very inexpensive and viable to run. Each child receives a meal prepared by volunteer mothers and by the children themselves.

Initially, there was some vandalism from local gangs, but with patience from staff and educators, and without police intervention, the gangs began to get involved in the programme. Among other things, PIA children are taught how to take care of younger children and how to clean and grow vegetables plus other skills which they can use elsewhere. Money the teenagers earn as gardeners is passed on to their *favela* neighbourhood association.

Before PIA, favela children were often isolated socially. Now they feel more part of a community. Family life has improved and the surrounding environment is being protected and improved instead of being destroyed. This programme has been named a United Nations Local Government Honours Programme from the International Council for Local Environmental Initiatives for 'environmental regeneration of low-income communities'. By the end of 1992 there were 28 PIA units operating (each catering for 250 children) and the programme has now been expanded to the whole state of Paraná.

In addition to launching environmental education programmes in low-income districts, the city has incorporated environmental education into its school curriculum. When Curitiba launched its recycling programme, city planners believed that the most effective way to teach the people about it was through the children. Much of the environmental education was taught by the 'leaf' family, actors dressed up as trees or leaves who visited schools, distributing brochures and who also appeared on television. The children responded positively to the idea of recycling (for instance collecting spent batteries and empty toothpaste tubes from their homes) and went home to teach their parents. Another important educational tool was the recent launching of the Free Open University for the Environment. This provides courses for people from all backgrounds (for instance taxi drivers, journalists, child carers) to encourage awareness of the environment and the importance of its preservation. It is also involved in research and in developing local environmental projects and is creating a data library.

Conclusions

• A city must know where it is growing, how and why. Conscious technical, political and economic decisions should be made in response to existing trends. Many urban-related problems linked to the uncontrolled physical expansion of cities can be avoided if correct decisions are made at the right time.

• There needs to be a close relationship between the public transport system, the land use legislation and the hierarchy of the urban road network.

• The most appropriate choice may represent a challenge to certain technological dogmas. Curitiba has showed that a city with more than one million inhabitants does not necessarily need a 'metro' style underground transport system or a light rail system and that surface solutions based on buses could be developed incrementally at a much lower cost. The city's solid waste programme has also shown that the recovery from household wastes of recyclable elements does not need an expensive mechanical separation plant, if a city transforms every household into a pre-separation plant with curbside collection schemes.

• City governments should understand the main economic opportunities and work towards developing them. The network of formal and informal economic relations should be supported and not hindered by urban planning actions.

• Total priority should be given to public transport rather than to private cars, and to pedestrians rather than to motorized vehicles. In Curitiba, less attention to meeting the needs of private motorized traffic has generated less use of private motorized traffic. Bicycle paths and pedestrian areas should be an integrated part of the road network and public transport system.

• A sustainable city is one that spends the minimum and saves the maximum. Hence the pragmatic application of recycling.

• Solutions within any city are not specific and isolated but inter-connected. Partnerships should be encouraged between private sector entrepreneurs, non-governmental organizations, public organizations, mixed capital institutions, neighbourhood associations, community groups and individuals. Taking this approach, the whole debate in favour of or against privatization loses its importance when we accept the simple fact that there is a role for each actor within a given community and city and that these roles can be complementary.

• The role of every actor is affected by scale, means and knowledge. For instance, the city administration should be in a position to determine structural guidelines for the city and its wider region whereas citizens can better determine what is better for their own street or neighbourhood. A good balance between representation and participation is essential.

• Creativity can substitute for financial resources. Ideally, cities should turn what are traditional sources of problems into resources. For instance, public transport, urban solid waste and unemployment are traditionally listed as problems but they have the potential to become generators of new resources and employment.

• Since information is essential, a team of officials should be developed locally who know the city well and who are committed to developing it. The better the inhabitants know their city, the better they treat it.

* A revised and updated version of an article which first appeared in *Environment and Urbanization, Vol 4 No 2 Oct 1992.*

The Contributors

Lucy Alexander is a journalist who specialises in Latin America. She became interested in community responses to Colombia's Pacific Plan while working in the press office of Christian Aid in London.

Charles Arthur is editor of *Haiti Briefing*, the newsletter of the London-based Haiti Support Group. He is author of *After the Dance, the Drum is Heavy*, HSG 1995, analysing Haiti in the aftermath of the 1994 US/UN intervention.

Anthony Bebbington is a geographer specializing in NGOs, popular organisations and the state. He has worked at the International Institute of the Environment and Development (London) and at the Social Policy Division of the World Bank. He is currently based at the University of Colorado. Publications include *NGOs and the State in Latin America: Rethinking Roles in Sustainable Agricultural Development*, Routledge 1993 (with Graham Thiele).

Nick Caistor is a senior producer at the BBC World Service. He is author of *Argentina in Focus*, Latin America Bureau 1996 and has translated several books, including *Born to Die in Medellín*, LAB 1992 and *Salsa*, LAB 1995.

Jorge Cela, S.J. is an anthropologist and Director of the P. Juan Montalvo Social Studies Centre in the Dominican Republic, working with popular organizations in Santo Domingo. He was the first Executive Director of Ciudad Alternativa (1988-92). He is author of *Una Ciudad con Participación Popular*, Ciudad Alternativa 1991 and co-author of *Educación y Cambio Social*, INTEC 1974, and *De Superman a Superbarrios*, CEAAL-CEASPA 1990.

Julio Dávila lectures at the Development Planning Unit, University College London, specializing in urban economic development, the urban environment, and gender issues. Previously he worked for Colombia's National Planning Department and the International Institute for Environment and Development (London and Buenos Aires). He is co-founder of the journal, *Environment and Urbanization.*

Elizabeth Dore is a senior lecturer in Latin American history at the University of Portsmouth, UK. She is author of *The Peruvian Mining Industry: Growth, Stagnation and Crisis*, Westview Press 1988, and several published articles on ecology. She is editor of *The Politics of Gender in Latin America* (forthcoming).

James Fair is a freelance journalist currently based in Ecuador. He is involved in a number of ecotourism projects in Ecuador's cloud forests. He has written for various planning and educational journals.

Al Gedicks is a professor of sociology at the University of Wisconsin-La Crosse and is a longtime environmental/native solidarity activist in the upper midwest of the USA. He has served as the director of the Center for Alternative Mining Development Policy and as the executive secretary of the Wisconsin Resources Protection Council.

Anthony Hall is a Senior Lecturer in Social Policy and Planning in Developing Countires at the London School of Economics and Political Science. Publications include *Developing Amazonia: Deforestation and Social Conflict in Brazil's Carajás Programme*, Manchester University Press, 1989/91 and (with David Goodman) *The Future of Amazonia: Destruction or Sustainable Development?* Macmillan 1990.

David Kaimowitz is a specialist at the Center for International Forestry Research in Jakarta, Indonesia. Previously he worked at the Inter-American Institute for Cooperation on Agriculture (IICA) in San José, Costa Rica. He has written widely on Latin America's agriculture and environment.

Judith Kimerling is a visiting scholar at Yale Law School and an expert on the environmental and social impacts of oil development and tropical forests. Previously she practiced environmental law in New York City and was an Assistant Attorney General for New York State. Since 1989 she has worked with various indigenous and settler organizations in the Ecuadorean Amazon. She is author of *Amazon Crude*, NRDC 1991, and numerous articles.

Catherine Matheson was trained as a journalist by the BBC and currently works for Christian Aid in London, specialising in development and economic issues. In 1984-86 she lived in El Salvador and covered Central America for the BBC, the *Guardian* and *Time* magazine.

Hilary McD. Beckles is Professor of Social and Economic History at the University of the West Indies in Barbados. He is author of several books on Caribbean history, including *Natural Rebels: A Social History of Enslaved Black Women in the West Indies*, Zed Books 1990, *A History of Barbados*, Cambridge, 1990, and editor of *Liberation Cricket: West Indies Cricket Culture*, Manchester University Press.

Marianne Meyn is a linguistics and political sciences graduate of Philipps University, Marburg, Germany. Since 1983 she has worked as a documentalist, fundraiser, and administrator at the Misión Industrial, Puerto Rico. Publications include studies of the colonial language question; articles on the US military in Puerto Rico; and various analyses of Puerto Rico's environment from an eco-social perspective.

Stephen Nugent is Senior Lecturer and Head of Department of Anthropology, Goldsmiths College, University of London, and Associate Fellow of the Institute of Latin American Studies, London. He has worked on Amazonia since the 1970s and is the author of *Big Mouth: the Amazon Speaks*, 4th Estate 1990 and *Amazonian Caboclo Society: an Essay on Invisibility and Peasant Economy*, Berg 1993.

Margarita Pacheco is an architect and urban planner. She is Assistant Professor at the Environmental Studies Institute of the National University of Colombia. She has worked as a consultant for various Colombian municipalities and for the Ministry of the Environment. She has published several articles on urban environmental issues and is the Colombian Coordinator of the Habitat II Best Practices award.

Polly Pattullo is a journalist on the *Guardian* and *Caribbean Insight* in London. She co-founded a tour-operating company specialising in the Eastern Caribbean and has written widely on the Caribbean region. She is author of *Last Resorts: the Cost of Tourism in the Caribbean*, Latin America Bureau 1996.

Aidan Rankin is Campaigns Press Officer for Survival International in London. Previously he taught Latin American Politics at the LSE. He has had articles published in *New Left Review* and *The Times Literary Supplement*.

Jonas Rabinovitch is Senior Urban Development Advisor for the United Nations Development Programme in New York. Previously he was Director of International Relations in Curitiba, Brazil, having joined the Curitiba Planning Institute as city planner in 1981.

Peter Rosset is the executive director of the Institute for Food and Development Policy (Food First) in Oakland, California. He is author of *The Greening of the Revolution: Cuba's Experiment with Organic Agriculture*, Ocean Press 1994 and *A Cautionary Tale: Failed US Development Policy in Central America*, Lynne Riemer 1996, among other books.

Denise Stanley is a Ph.D Candidate in Agricultural Economics at the University of Wisconsin, Madison. She has worked in Honduras and the Dominican Republic with community development and environmental organizations. She is the author of several articles on Central American natural resource exports.

Sarah Stewart is a journalist with Christian Aid in London. She is the author of numerous articles on Latin America and co-author of two books, *South Africa, Inc*, Simon & Schuster and Yale University Press 1986, on the Anglo American Corporation, and *Sweet Ramparts*, War on Want 1983, on women in Nicaragua.

Lori Ann Thrupp is Director of Sustainable Agriculture at the World Resources Institute's Center for International Development and Environment, Washington DC. She is author of *Bittersweet Harvests for Global Supermarkets: Challenges in Latin America's Agricultural Export Boom*, WRI Publications 1995.

Janet Townsend is a senior lecturer in Geography at the University of Durham, UK, specialising in gender, ecopolitics and rural change in poorer countries. She has thirty years' intermittent experience of working with pioneers in the rainforest, particularly in Mexico and Colombia.

Index

Also published by

WOLLASTON

People Resisting Genocide

Miles Goldstick

Foreword by Dr. Rosalie Bertell

Natives' struggle in northern Saskatchewan to protect their homes from the effects of uranium mining.

These are important issues, and in raising them Goldstick does us a service.
Border/Lines

315 pages, photographs, illustrations
Paperback ISBN: 0-920057-95-0 $16.99
Hardcover ISBN: 0-920057-94-2 $45.99

MASK OF DEMOCRACY*

Labour Suppression in Mexico Today

Dan LaBotz

an ILRERF Book

Following scores of interviews with Mexican workers, labour union officials, women's organizations, lawyers, and human rights activists, Dan LaBotz presents this study of the suppression of workers' rights in Mexico.

Now North America can see why the situation for Mexican workers continues to get worse.
Mexican Acton Network on Free Trade

223 pages, index
Paperback ISBN: 1-895431-58-1 $19.99
Hardcover ISBN: 1-895431-59-X $48.99

ELECTRIC RIVERS

The Story of the James Bay Project

Sean McCutcheon

... a book about how and why the James Bay project is being built, how it works, the consequences its building will have for people and for the environment... it cuts through the rhetoric so frequently found in the debate.
Canadian Book Review Annual

194 pages, maps
Paperback ISBN: 1-895431-18-2 $18.99
Hardcover ISBN: 1-895431-19-0 $47.99

DYING FROM DIOXIN*

A Citizen's Guide to Reclaiming Our Health and Rebuilding Democracy

Lois Marie Gibbs

It's in our food. It's in our water. It's in our bodies. And it's making us sick. According to recently released studies, widespread exposure to dioxin is destroying the health of the people. In *Dying from Dioxin*, Lois Marie Gibbs and grassroots activists at the Citizen's Clearinghouse for Hazardous Waste describe the alarming details of this public crisis, and explain how citizens can organize against this toxic threat.

200 pages, index
Paperback ISBN: 1-55164-084-8 $19.99
Hardcover ISBN: 1-55164-085-6 $48.99

TRIUMPH OF THE MARKET*

Essays On Economics, Politics, and the Media

Edward S. Herman

Who lubricates the machine that controls trade and finance on a global scale? What force keeps us money-hungry and always waiting for the next sale? In this book, Edward S. Herman reveals the power-magnet market operators who dictate how economic life is organized the world over.

A disturbingly blunt warning about the clear and present dangers to democracy, economic rationality, national sovereignty, global economic stability and progress, and international peace.
Samori Marksman, WBAI-FM, Pacifica Radio

286 pages, index
Paperback ISBN 1-55164-062-7 $19.99
Hardcover ISBN 1-55164-063-5 $48.99

POLITICS OF SUSTAINABLE DEVELOPMENT

Citizens, Unions and the Corporations

Laurie E. Adkin

The attitudes and actions of citizen's groups, unions and corporations reflect not only their stakes in protecting particular interests, but also the limits of their abilities to envision, or mobilize support for, alternatives to the prevailing mode of economic growth. Growing public concern about toxic chemical and industrial pollution issues coincide with a peak in environmental activism and government initiatives. These challenges are then set alongside the complex problem of labour movement responses.

250 pages, photos, maps, illustrations
Paperback ISBN 1-55164-080-5 $19.99
Hardcover ISBN 1-55164-081-3 $48.99

MYTH OF THE MARKET

Promises and Illusions

Jeremy Seabrook

The majority of the people place their hope and faith in the mechanism of the market as the bearer of promise for the future, but the spreading of market values leads to social disintegration and the destruction of cultures.

A strong indictment of the market system. All the more timely with the recent moves in global trade.
Peace and Environment News
There are alternatives to the market, but unless we begin to resist the monetization of all human activity, they will be relegated to museums where, fittingly, you'll pay to see them.
Imprint

189 pages
Paperback ISBN: 1-895431-08-5 $18.99
Hardcover ISBN: 1-895431-09-3 $47.99

COMPLICITY

Human Rights and Canadian Foreign Policy: The Case of East Timor

Sharon Scharfe

1995 marks twenty years since Indonesia invaded the country of East Timor; in that time, over one-third of the population has been killed. The atrocities continue, assisted either directly or indirectly, by Western governments. This in spite of several resolutions at the United Nations condemning the illegal invasion of East Timor and calling for Indonesia's immediate withdrawal.

This book concludes that while Canada has occasionally shown signs of concern, its foreign policy remains "business-as-usual." In failing to take a strong stance on the gross and systematic human rights violations ongoing in East Timor, the Canadian government emerges as an accomplice in the bloodshed.

This book is designed to keep information flowing.
Windsor Star

Sharon Scharfe works with the East Timor Alert Network, and runs the International Secretariat for Parliamentarians of East Timor. She holds an MA in legal studies from Carleton University.

Photographs by Elaine Brière

249 pages, index, 15 photographs
Paperback ISBN: 1-55164-032-5 $19.99
Hardcover ISBN: 1-55164-033-3 $48.99
L.C. No. 95-79351

COMMUNITY ECONOMIC DEVELOPMENT

In Search of Empowerment and Alternatives

Eric Shragge, ed.

Challenges the notion that the economy should only be privately owned and argues that it should both act in the social interest of the local community and be partially controlled by it.

A critical discussion of both the theory and practice of community economic development.
Journal of Economic Literature

141 pages
Paperback ISBN: 1-895431-86-7 $19.99
Hardcover ISBN: 1-895431-87-5 $48.99
L.C. No. 93-072747
ISSN: 1195-1850

THE POLITICS OF URBAN LIBERATION

Stephen Schecter

A wide-ranging libertarian evaluation dealing with political economy in an urban context, from France to Chile. it also examines the importance of the city in the history of social revolution.

203 pages
Paperback ISBN: 0-919618-78-2 $9.99
Hardcover ISBN: 0-919618-79-0 $38.99

BUREAUCRACY AND COMMUNITY

Linda Davies, Eric Shragge, eds.

This book examines the consequences for both State social workers and community practitioners in face of increasing governmental restraints.

Based on recent empirical work from Québec and the United Kingdom.

... takes a highly critical view of social-services management and the controlling role of government bureaucracies.
Calgary Herald

180 pages, bibliography
Paperback ISBN: 0-921689-56-X $16.99
Hardcover ISBN: 0-921689-57-8 $45.99
L.C. No. 90-81638
ISSN: 1195-1850

YEAR 501

The Conquest Continues

Noam Chomsky

2nd printing

A powerful and comprehensive discussion of the incredible injustices hidden in our history.

...Year 501 *offers a savage critique of the new world order.*
MacLean's Magazine
Tough, didactic, [Chomsky] skins back the lies of those who make decisions.
Globe and Mail
... a much-needed defense against the mind-numbing free market rhetoric.
Latin America Connexions

331 pages, index

Paperback ISBN: 1-895431-62-X	$19.99
Hardcover ISBN: 1-895431-63-8	$48.99

JFK, the Vietnam War, and U.S. Political Culture

Noam Chomsky

For those who turn to Hollywood for history, and confuse creative license with fact, Chomsky proffers an arresting reminder that historical narrative rarely fits neatly into a feature film.

... a fascinating and disturbing portrait of the Kennedy dynasty.
Briarpatch
... the most important contribution to the ongoing public and private discussions about JFK.
Kitchener-Waterloo Record

172 pages, index

Paperback ISBN: 1-895431-72-7	$19.99
Hardcover ISBN: 1-895431-73-5	$48.99

ANARCHIST COLLECTIVES

Workers' Self-Management in Spain 1936-39

Sam Dolgoff, ed.

Introduction by Murray Bookchin

The eyewitness reports and commentary presented in this highly important study reveal a different understanding of the nature of socialism and the means for achieving it.
Noam Chomsky

195 pages, index, bibliography

Paperback ISBN: 0-919618-20-0	$16.99
Hardcover ISBN: 0-919618-21-9	$45.99

RACE, GENDER AND WORK

A Multi-Cultural Economic History of Women in the United States

Teresa Amott and Julie Matthaei

Race, Gender, and Work *is exciting because of its frank acknowledgement of difference among women. It is a volume that will inform and motivate scholars and activists.*
Julianne Malveaux, University of California, Berkeley
... a detailed, richly textured history of American working women.
Barbara Ehrenreich, author of The Worst Years of Our Lives

433 pages, index, appendices
Paperback ISBN: 0-921689-90-X $19.99
Hardcover ISBN: 0-921689-91-8 $48.99

BALANCE: ART AND NATURE

John Grande

Poses questions about our relationship to the natural world, our place in the stream of natural evolution and technological process, and the role of visual artists in understanding, and defining, society and nature.

Makes unexpected connections giving new insights into contemporary art.
Public Art Review
Grande's book contains a lot of ideas, all of which are thought-provoking.
Globe and Mail
Grande's grasp of the details makes this book convincing.
Books In Canada
Offers interesting parallels between different aspects of public art.
Espace Sculptur

250 pages, photographs, index
Paperback ISBN: 1-551640-06-6 $19.99
Hardcover ISBN: 1-551640-07-4 $48.99

POLITICAL ECONOMY OF INTERNATIONAL LABOUR MIGRATION*

Hassan Gardezi

While former studies on labour migration have concentrated on its effect on GNP, foreign exchange earnings, and labour exporting countries' rates of investment, Gardezi's work refocuses attention on the migrant workers themselves, their hopes and aspirations, family and community life, and working conditions both at home and abroad. Taking this wide-ranging view, he is able to enhance our understanding of the transfers of labour force.

210 pages
Paperback ISBN: 1-551640-16-3 $19.99
Hardcover ISBN: 1-551640-17-1 $48.99